KU-655-745

The Way Home

Making Heaven on Earth

Madis Senner

LIBRARIES NI
WITHDRAWN FROM STOCK

BOOKS

Winchester, UK
Washington, USA

CONTENTS

Preface

In 1997 my world changed. I left Wall Street where I had worked the previous 15 years and my wife of 10 years and I decided to divorce. Unbeknownst to me I was beginning a spiritual quest. While I have always been spiritually attuned it seemed that my world began expanding to include other unseen worlds.

I started reading everything I could to learn more about spirituality and the mystical from scripture to books on the occult. I took up meditation and other spiritual practices to develop my sixth sense. Gradually I began to get a sense, through inner knowing as well as a voice within me, that I was being called to be a servant of God. This meant that I was to give up deciding the direction of my life and let God guide me.

One evening in February of 1999 I accepted God's offer and committed my life to God. I felt a strong confirming chill within me as well as a sense of joy and purpose. As I gazed out upon the New York skyline I felt that God was right there with me sharing a beautiful dream. No words were spoken, yet I felt God was communicating to me in some way about the wondrous world to come and how I was going to help bring heaven on earth. It sounded like a fine and noble idea, but what was heaven upon earth?

It was several years before I had the remotest idea of what heaven on earth was about. I would be led in many directions. I left New York City and moved back to the home where I grew up in Syracuse, New York. I became a social activist and eventually came to the aid of a local Muslim doctor, Rafil Dhafir, who was targeted and selectively prosecuted by the Bush administration for providing humanitarian aid to Iraq in defiance of the law in the 1990's.

In the summer of 2002 I felt called to get people to pray around Onondaga Lake in Syracuse, NY. That began a big refocus and learning process that would blossom into www.MotherEarthPrayers.org and the cataloguing of sacred sites in upstate NY. I would come to learn about Mother Earth's spiritual dimension and our dynamic relationship with her. Through my focus on Mother Earth I would gain sentience of her and come to understand how our thoughts affect her and us.

In the summer of 2007 I wrote a letter to a friend asking the questions I had been pondering about, the direction of the world and the shape of things to come. Over the course of the summer those seed thoughts about our future direction would bear fruit and blossom in me, planting many more seed thoughts. By that fall I knew it was time to write a book about bringing heaven upon earth.

What follows is what I have learned these past few years.

I would like to thank all those that have helped in my spiritual path, beginning with Mother Angela for helping me keep at least one foot in this world and being a friend for listening to what must have been blathering at times. Thanks to Bill Attride for giving me direction and helping me understand what I was to learn from the life lessons I was presented as I progressed on my path. Thanks to Tony Bartlett of the Wood Hath Hope bible community for teaching me about the difference between a God of Love and a God of Wrath. Special thanks to Lorraine Mavins for her counsel and editorial advice and for her friendship in traipsing around New York State looking for and visiting sacred sites.

Madis Senner
www.MotherEarthPrayers.org

Introduction

The belief in an idyllic life whether it is in this world or in the life to come is universal to all peoples of the earth. This belief in a better existence is inherent in us and drives us. For some this means working hard to acquire more things. Some work hard to make ends meet and it means dreaming of a better life for their children and the life after. For others it means living a spiritual life so that they may advance to heaven. This longing and questing is a tacit acknowledgement that there is something finer beyond our current existence. It is an affirmation that what we have is not complete.

Scripture offers the hope of a paradise and a better life to come. Buddhists and Hindu's look to live a good and moral life so that they may be rewarded with a better life in their next re-birth on their ladder to Nirvana[1], or Moksha[2]. Similarly the Abrahamic traditions believe that righteous living will be rewarded with a life after in heaven. They all believe that ultimately a better life is to be found beyond the physical world; that a paradise exists in some sort of mystical world beyond the physical world.

If the earth is not where the mystical paradise we wish to escape to resides, what is the earth?

That is an important question to ask since it appears we care little about Mother Earth and our relationship with her. Manmade global warming threatens the very existence of the earth. The twentieth century was the most violent century in the history of civilization with more people having died in war than from all the previous wars put together. Globalization has usurped democracy as multinational corporations rule the world. Most certainly we have turned earth into a hell, or a dystopia of sorts.

Scripture tells that earth holds the possibility of being a

paradise. In the Hebrew creation story of the Garden of Eden in Genesis, paradise was located on earth where a large river separated into four branches.[3] The Garden of Eden speaks to a vision of living in harmony with God and nature. Similarly God tells Moses that he is taking the Israelites to the Promised Land of milk and honey.[4] Isaiah holds out the hope of a time of peace, tranquility and love between all peoples and creatures:

> The wolf shall live with the lamb, the leopard shall lie down with the kid, the calf and the lion and the fatling together, and a little child shall lead them... They will not hurt or destroy on all my holy mountain; for the earth will be full of the knowledge of the LORD as the waters cover the sea.[5]

The hope of a Promised Land, a piece of heaven on earth, has led many to seek alternative lifestyles and to live outside of the mainstream. Religious communities have sprung up through out history as have modern Utopian communities. All of these enclaves were established to better live in accordance with God.

The fact is that earth holds out hope of containing a paradise. The earth is meant to be a paradise, a heaven of sorts. In the Lord's Prayer, the core prayer of the Christian faith, Jesus tells us to pray that earth be like heaven: "Thy kingdom come. Thy will be done, on earth as it is in heaven."[6]

By telling us that God's will shall be done on earth as it is heaven, Jesus is saying that earth shall be no different than heaven, meaning that earth should be transformed into heaven. Jesus goes on to describe how the worship and love of God along with right living, forgiving and loving others, not hoarding material possessions, not worrying, will make heaven upon earth.[7]

Instead of making heaven upon earth, we have killed, raped and pillaged each other and Mother Earth. We have turned earth into a hell dominated by the unrighteous. Jesus reaffirms this in the beatitudes when he turns the conventional wisdom of the ruling power structure upside down by blessing those that have suffered under it and been its losers. By blessing the downtrodden Jesus was saying that those prospering in the system were the true losers in the kingdom of heaven. The winners in the unjust system have only gotten larger and more hideous since the time of Jesus.

It would be a mistake to assume that the unrighteous power structures dominating the world at the time of Jesus and today are representative of what the earth is truly meant to be. They are not. They are like a cancerous growth that must disappear if we are to heal and live a good and healthy life. In the third beatitude Jesus says, "Blessed are the meek, for they will inherit the earth."[8] In other words heaven will be upon earth when love, gentleness and humility reside in the consciousness of humanity; a tall order.

Making Heaven upon Earth

The task of making heaven upon earth begins by taking to heart what we have been taught throughout the ages and applying it in a different manner.

The words of Buddha, Jesus, Krishna, Mohammed, Moses, Mahavira and the Peacemaker resonate today as they did back then. Their teachings are timeless and they can help guide us to the Promised Land here on earth.

What is different is that we need to realize that we live in a supernatural world and act accordingly. If earth is to be like heaven and heaven is a mystical paradise, then earth is to be a mystical paradise as well. Earth is a metaphysical paradise, but what is missing is our lack of righteous living.

Living and functioning in an unseen world is very different from living in a physical world. It means living beyond the world of physical senses and being aware of all that we think and do and how this reverberates. For example, we might be kind to someone we meet but at the same time think bad things about them. In the physical world such a meeting would be considered a positive event. In the ethereal world it would be just the opposite as the negativity of our thinking would dominate the meeting, as would the thoughts of the other person. Those negative thoughts also have a potential to become something much greater than they are.

What transpires in the unseen world influences and often dominates what happens in the physical world. The rules of the mystical world, while very similar to the physical world, can appear to be quite different at times. The transcendental world is also filled with all sorts of inhabitants and beings, both good and bad. It is a world we know little about, yet is the foundation of our existence.

To make heaven upon earth, or transform the earth into heaven, we will need to learn how to work with the metaphysical world. Not only is transformation of the earth possible, it does not require the mass of humanity to be involved. Because transcendental knowledge is empowering a few people acting with that knowledge can have a tremendous influence. The focus of this book is to educate people about that metaphysical knowledge so that we can liberate ourselves and bring about heaven on earth.

Part I the Mystical World, details the mystical world and its various realms and their inhabitants. It explains how our thoughts once created exist independently of us in the unseen world and how they exert a great influence upon us. You will learn about Mother Earth's cosmology and how ideally we dynamically interact with her in a symbiotic and nurturing relationship. If we are to solve the problems of global warming and pollution we must learn that the same mind body dynamic at work within us is also at work within Mother Earth, only it is our thoughts and actions that influence her body. The solution to healing the world rests in our hearts not in our minds.

Part II Hell: Our World, Our Making, explains how by ignoring the ethereal world we have trapped ourselves in a hell of our own making. Our collective thoughts have created a complex unseen world ruled by idols and false gods that are constantly looking to manipulate us. Much of what we hold dear and cling to is an illusion that takes us farther away from our true nature.

Part III Heaven on Earth, provides a road map for liberating ourselves and saving the planet. That which has trapped us can also free us; if our collective thinking has ensnared us in the physical world it can also free us. The final part of the book will show how we can affect great change if we apply our thoughts and actions in a focused manner. Small groups of people working with knowledge of the mystical world and Mother Earth can have an exponential influence. If we are dedicated and diligent we can heal Mother Earth and bring about peace and happiness. We can make heaven upon earth a reality.

The path that I am asking you to walk is not the path of individual salvation, but the path of our collective liberation.

Part 1 The Mystical World

Chapter 1

Hidden World(s)

We live in a mystical world.

There are worlds around us that we cannot see, worlds that are beyond our physical senses. We dynamically interact with these worlds and they are constantly influencing us. They are filled with our past thoughts and actions, other beings and their thoughts and actions and much, much more. These worlds are part of our existence yet we don't perceive them because we have plunged headlong into the physical world and blinded ourselves to the true reality of our existence. By focusing on the physical world we have come to identify with it and in doing so, have lost touch with our true spiritual self.

Sri Aurobindo says that,

> [W]e are not alone in the world; the sharp separateness of our ego was no more than a strong imposition and delusion; we do not exist in ourselves, we do not really live apart in an inner privacy or solitude. Our mind is a receiving, developing and modifying machine into which there is being constantly passed from moment to moment a ceaseless foreign flux, a streaming mass of disparate materials from above, from below, from outside. Much more than half our thoughts and feelings are not our own in the sense that they take form out of ourselves.[1]

To comprehend the hidden world we need to look beyond our physical senses and into the supernatural. That means relying on what many call our sixth sense.[2] All of us have a sixth sense to a varying degree. Unfortunately we have become so attached to the physical world and the pleasure senses that we have lost touch with and have not developed our sixth sense. Reclaiming our sixth sense begins with spiritual exercises and the desire to experience God. It also means refocusing our life from the physical to the spiritual. It is only when we have developed our sixth sense and refocused ourselves on the unseen world and God that we will be able to experience these other worlds.

We are inherently spiritual beings that have lost our true selves and have become entangled in the world of sense objects. Trappist monk Thomas Merton says that the fall from paradise by Adam and Eve was the fall from unity:

> St. Augustine, in a more cautious and psychological application of the narrative, says that in the Fall Adam, man's interior and spiritual self, his contemplative self, was led astray by Eve, his exterior, material self, his active self. Man fell from the unity of his contemplative vision into the multiplicity, complication, and distraction of an active, worldly existence. Since he was now dependent entirely on exterior and contingent things, he became an exile in a world of objects, each one capable of deluding and enslaving him.[3]

We are spiritual beings having a physical world experience.

Maya

The physical world we cling to and live in is not real. Hindu scripture says that the physical world is an illusion created by Brahman (God) and behind the illusion is God: "The whole world is filled up with his (God's) members."[4] The visual deception of the physical world is carried out by maya[5] the power of illusion that veils all in the physical world. Because of maya we do not see the true nature and reality of the world.

Hindus believe that it is maya that binds and blinds us to the physical world. By deceiving us into believing that the physical world is real, it draws us ever more closely into it. In the physical world we lose track of our true being and develop attachment (like, dislike, desire, love, hate, etc.) to physical world and sense objects.

Buddha in the Dhammapada (the Path of Dharma) similarly tells us that the world is an illusion and that we should, "Look upon the world as a bubble, look upon it as a mirage."[6]

Michael Talbot in *The Holographic Universe* describes the illusion of physical reality as being like a giant hologram that we see and experience. He compares it to the beginning of the movie Star Wars when Artoo Deetoo projects a three-dimensional image of Princess Leia: "The image is a hologram, a three-dimensional picture made with the aid of laser, and the technological magic required to make such images is remarkable. But what is even more astounding is that some scientists are beginning to believe the universe itself is a kind of giant hologram, a splendidly detailed illusion."[7]

Anyone who has meditated for any time has experienced the realization that the physical world can be blocked out and that there is something greater. Deeper states of meditation lead to the temporary loss of the physical senses. Accomplished yogis in deep trance states, where they have broken the link with the physical world, have been able to endure all sorts of physical pain without flinching. Tibetan Buddhist monks practicing the meditative technique of Tummo are able to withstand 40′ Fahrenheit temperatures with cold wet towels draped over their bare backs and not be affected. Most people would be shivering uncontrollably or even die. Herbert Benson who studies the benefits of meditation and other techniques notes that the monks were able to overcome their physical senses because they understood that the physical world is not real: "Buddhists feel the reality we live in is not the ultimate one. There's another reality we can tap into that's unaffected by our emotions, by our everyday world. Buddhists believe this state of mind can be achieved by doing good for others and by meditation. The heat they generate during the process is just a by-product of g Tummo meditation."[8]

If the physical world is an illusion where is this other ultimate reality, the one in which the Tibetan monks are able to transcend the physical world? Jesus gives us a clue when he talks about the kingdom of heaven: "The kingdom of God is not coming with things that can be observed; nor will they say, 'Look, here it is!' or 'There it is!' For, in fact, the kingdom of God is within you."[9]

The idea that heaven, or the ultimate reality, cannot be seen and is found within ourselves, is similar to the beliefs of the eastern traditions. The idea that heaven exists inside of ourselves is as much about where heaven is as it is about the direction we should look to find it.

All is God, God is All

Hindu Vedanta holds that the reality of what our senses are experiencing in the physical world is Brahman, the supreme spirit, the absolute reality, in many ways what other's call God. The concept that Brahman is all, the ultimate reality, is reiterated throughout Hindu scripture,[10]

Just as, my dear, bees prepare honey by collecting the essences of different trees and reducing them into one essence. And, as these 'juices' posses no discrimination (so that they might say), 'I am the honey of this tree, I am the honey of that tree,' all creatures though they are composed of the same essence, do not realize that they are. They see themselves as whatever they are in this world, tiger, lion, wolf, boar, fly, or gnat, worm, or mosquito. That which is the subtle essence is the same for this wormhole. That is the truth. That are thou.[11]

The *Essene Gospel of Peace (Book I)* speaks similarly of our divine

11

connection to each other and the world around us. In it Jesus emphasizes our divine connection to the earth:

> And Jesus answered…I tell you truly, you are one with the Earthly Mother; she is in you, and you in her. Of her were you born, in her do you live, and to her shall you return again. Keep, therefore, her laws, for none can live long, neither be happy, but he who honors his Earthly Mother and does her laws. For your breath is her breath; your blood her blood; your bone her bone; your flesh her flesh; your bowels her bowels; your eyes and your ears are her eyes and her ears.[12]

Unfortunately, the reality that Mother Earth and our relationship to her are integral for our survival is the antithesis of classic thought and even heresy in some organized religions.

Science is increasingly coming to the realization of the unity and oneness of the universe as physicist Fritjof Capra tells us: "The basic oneness of the universe is not only the central characteristic of the mystical experience, but is also one of the most important revalidations of modern physics."[13]

As it is for one, So it is for Many

The notion that all is Brahman and that we are of one essence has many important implications. It means that we are linked together and are part of a greater union, God. It also means Mother Earth, Gaia, is part of Brahman and hence part of the unity, as are all of God's creatures.

Since we are all part of Brahman we have similar compositions. While there are differences that can be significant, ultimately in the larger sense, we have similar make-ups. As we shall later discuss, the functioning of our subtle body is very similar to that of Mother Earth's.

Because we are of one essence and part of God what applies to each of us individually applies to all of us collectively. As it is for one, so it is for many, since many are one and one is many. The Chândogya Upanishad says a very similar thing about God manifesting in the physical world as many; "may I be many, may I grow forth."[14]

So the laws of the universe apply to us individually and collectively.

Planes of Existence

The apparent duality between reality (Brahman) and illusion (the physical world) is further expounded in the belief that world is made up of purusha (consciousness) and prakriti (energy) and combinations thereof.[15] Purusha is pure consciousness and prakriti is energy—nature, matter and all that is known and seen by the physical senses. Everything around us is made up of consciousness (purusha) and energy (prakriti) to varying degrees. Our soul is pure consciousness, all else is a mix. Inanimate physical objects are almost pure energy (prakriti).

While the universe is made up of purusha and prakriti there are structures, configurations, multiple dimensions and unions that create a very complex order beyond the physical world. There are many invisible worlds, or realms that simultaneously exist with the physical world. Each of these realms is a separate world unto itself, with separate inhabitants and structures, but at the same time they overlap each other. Many refer to these worlds as planes of existence, others describe them as other dimensions; the Hindu's refer to them as Lokas, or spheres. The Tibetan Buddhists practicing Tummo were able to escape the reality of cold temperatures by having their consciousness retreat to one of these planes of existence.

We know little of the various planes of existence. Mystics have given some insight, but there is no consensus of opinion. There are also sub-planes that create a very complex cosmology, the number and nature of varying between different beliefs systems. The Mundaka Upanishad says that there are seven worlds[16] similarly the Hindu Purana holds that there are what it calls seven lokas.[17] Hindu tantrics believe that there are seven "sapta bhumikas," or levels of existence, each one like a ladder which we must climb to attain higher states of consciousness.[18] Certain Buddhist sects believe in 31 planes of existence.[19] Theosophist Madame Blavatsky believed in seven planes of existence and numerous sub-planes.[20] Kabbalahists, Sufis and many others similarly believe in other planes. Even science's string theory holds the possible existence of several dimensions or universes.

There is uniformity of opinion that there is a hierarchy to the planes with the physical plane being the lowest. Swami Vivekananda says that the higher planes are associated with a progressive diminishing of the ego and the sense of 'I' and that superconsciousness or Samadhi is only achieved when our consciousness has transcended the middle plane of 'I'.[21]

The Physical Plane

The physical plane is what most of us call reality. But it is an illusion, a false reality that is created and maintained by apparatus in the pranic plane. It is the lowest and coarsest plane and is where our physical being and the material world reside. It is the world of the senses, of pleasures and of objects that can titillate us and, if we let them, they will drag us deeper into the quagmire of the physical plane. It is full of snares. Buddha taught us that it is attachment to the physical world that causes suffering.[22] Implicit in all faiths is the belief of something greater beyond the physical plane, heaven, the after-life and bliss.

The physical world contains that which is manifested. What distinguishes what we see and experience through our senses in the physical plane from what hovers around us is manifestation. The Bhagavad-Gita notes this in talking about the eternal soul and how life is manifest and death is un-manifest: "All created beings are unmanifest in their beginning, manifest in their interim state, and unmanifest again when they are annihilated."[23]

Consciousness is the highest force, or ideal, in the physical plane. It influences everything from our thoughts and actions to our subtle body to Mother Earth. The physical plane is greatly influenced by the mental plane of our thoughts looking to manifest as will be discussed in the next chapter. The physical plane is our school, only much more. We are here to learn and develop. There are certain aspects, or rules that govern the physical plane. Understanding how the rules work and how to work with them is the path to freedom as taught by the prophets and mystics of all faiths.

The Law of Karma

Karma, or the law of justice, operates in the physical plane. Madame Blavatsky calls it the Ultimate Law of the Universe: "Karma is that unseen and unknown law which adjusts wisely, intelligently and equitably each effect to its cause, tracing the latter back to its producer."[24] In other words, there are consequences for our actions. Whatever you do, for either the good or the bad, will come back to you like a boomerang from someone else. So if you give something away, something will be given back to you, or if you hurt someone, you will be hurt. Karma insures that we live with the consequences of our decisions and that everything ultimately balances. Unfortunately, when we look at the world we see injustice and imbalances all over. The poor are trampled, the rich grow richer and violence abounds. The balancing of karma takes several reincarnations of the participants' lives to manifest. Secondly, it would be wrong to make judgments about winners and losers because it is very much more complex. For example, someone could be taking on someone else's karma or have chosen a particular path that brought hardship so that they may grow. It is wrong to look at someone suffering and say that they are being paid back their karma. The winners or "successful people" in the physical world are not necessarily the winners in the spiritual world.

Group Karma

Unity and oneness means that not only do we have a responsibility to help others but we are to a certain degree responsible for each other. In other words if someone is doing good things we all benefit from them; conversely if they are doing bad things, we all build bad karma. We are called to intercede.

We share a common existence and unity, all is Brahman. While we may spiritually advance individually our ultimate progress is contingent upon the advance of everyone. Because we are of the "same essence" the right hand reaching to heaven cannot get there if the left is in the gutter of violence.

Madame Blavatsky speaks of group karma and how we can be swept up by the collective actions of others with whom we are united by community, religion, nation and the world. She speaks of what she calls "Distributive Karma" and that no one can rise up without helping others,

> that no man can rise superior to his individual failings, without lifting, be it ever so little, the whole body of which he is an integral part. In the same way, no one can sin, nor suffer the effects of sin, alone. In reality, there is no such thing as 'Separateness'; and the nearest approach to that selfish state, which the laws of life permit, is in the intent or motive.[25]

Group karma can be seen with the Hebrew prophets who continually reminded the people that there was a price to be paid for defying and defiling God. The prophet Hosea said that the house of Jehu was to suffer for wiping out Ahab's descendants in Jezreel: "[F]or in a little while I will punish the house of Jehu for the blood of Jezreel, and I will put an end to the kingdom of the house of Israel. On that day I will break the bow of Israel in the valley of Jezreel."[26]

Jehu's descendants were subjected to the group karma of murder and violence for the military might he exercised in the valley of Jezreel. Some may argue that this shows God to be violent. In reality it is the law of karma at work. Jehu did not have God's blessing to kill, because God is not for violence, ever, as this example shows and Hosea continually harped about. Group karma means that our fates are intertwined.

Law of Love

Love is God's highest aspiration for us. We are called to love one another, to love our enemy, to turn the cheek and to love others regardless of whether they love us in return. This is the law of love. It is the barometer by which our spiritual evolution is gauged.

All faiths have at their core the law of love. The Jewish tradition says it well. Rabbi Hillel when asked to explain Judaism to a man standing on one foot said, "What is hurtful to you, do not do unto others. Now go and study the rest."[27]

Transformation

The physical world around us is, or at least it appears to be, in a constant state of flux. Change is one of the features of the physical plane that affects us in so many ways. We grow, we age and we die. There is no escaping change because of the cycle of life and death.

All things continuously evolve from God's creatures to the world around us. The purpose of change in the physical plane is to help with our spiritual evolution. We are here to raise our consciousness by learning to love and by diminishing our ego and our sense of self. Everyone and everything is here to learn and evolve.

Jesus tells us that we must be reborn in spirit, in God: "'Very truly, I tell you, no one can see the kingdom of God without being born from above."[28]

The idea of spiritual rebirth does not end with accepting God, we are continually being reborn and evolving as we get closer to God and develop divine qualities. When we become rigid and inflexible, whether it is in refusing to accept new spiritual alternatives or with our physical body, we are not growing and are decaying.

The challenge is to remain firm but flexible at the same time. While it is good to be committed and dedicated to our beliefs, if we are not open to new vistas we are not advancing. In my spiritual transformation I have constantly found myself challenged to accept what is being revealed to me. For example, I was more than a little fraught when I felt called to get people to pray on the shores of Onondaga Lake in Syracuse, NY. The idea that a specific location, the land, could help people and the earth heal seemed surreal. Fortunately I persevered and that began my genesis in learning about Mother Earth.

We foolishly try to circumvent change and the cycle of death and life, seeking immortality for ourselves and for entities such as corporations, religious organizations and institutions. While aging brings knowledge, a healthy perspective and many more benefits, as we get older our mind and body eventually begin to whither and ultimately we die. No one escapes the disease, physical deterioration and dementia that old age brings if we live long enough. Yet somehow we give immortality to institutions thinking that they are impervious to the laws of the physical plane. The degenerative affects of aging are all too apparent in the behavior of corporations, religious organizations and institutions in the form of the abuse of people, the exploitation of the environment and abusing the political system for personal gain.

Soul Development

Our soul is eternal and does not die with the death of the body, but goes through a series of rebirths and reincarnations in the physical plane .[29] The purpose of being reborn is to spiritually mature. We progress through good thoughts and actions and devolve with bad thoughts or actions of selfishness, hate and violence. Swami Vivekananda notes how each deed we do affects our soul:

> Every wicked deed contracts the nature of the soul, and every good deed expands it; and these souls are all part of God. 'As from a blazing fire fly millions of sparks of the same nature, even so from this infinite being, God, these souls have come.' Each has the same goal...Souls are limited, they are not omnipresent. When they get expansion of their powers and become perfect, there is no more birth and death for them; they live with God for ever.[30]

So if we are thinking and doing violent acts or focusing on material possessions, we are curbing, if not regressing, our spiritual development and diminishing our soul. Spiritual exercises like meditation, praying, and contemplation can help with our spiritual development but ultimately it is what we think and do that determine our progression. Later Part III it will be discussed how spiritual exercises such as meditation can raise your consciousness and improve our ability to love, be compassionate and non-violent.

The development of the soul is the spiritual manifestation of our evolution and transformation. It is the gauge of our development. The sensitive person can sense an evolved soul. Transformation of the soul begins with how we treat others and then ourselves.

The Pranic Plane, Plane of Energy

The pranic plane is located immediately above the physical plane. It is the plane of energy and is responsible for regulating prana (chi, life force) and maintaining the physical plane. It contains all of the apparatus necessary to regulate the flow of prana. It acts like a large electrical or heating system for our home, only its function is to regulate the physical plane. It is also the plane of emotions and desires.

Prana is the life force that sustains us and powers us like electricity lights a bulb. We cannot live without prana. The heart may beat but it is prana that gives us life. Swami Niranjanananda Saraswati says that "As long as prana is retained, the body will not die…. Without prana we would be decaying corpses with no ability to see, move or hear."[31]

What I call the pranic plane is made up of all the sub-planes that are all involved with energy and maintaining the physical plane and correspond with our pranayama kosha as I will shortly discuss. From a Theosophist's perspective the pranic plane would consist of the etheric plane[32] and astral planes together. The astral plane[33] is the plane of emotions and desires. C. W. Leadbeater tells us that the astral plane is the first place we go to when we die: "[B]ut what we have now to consider is the lower part of this unseen world, the state into which man enters immediately after death – the Hades or underworld of the Greeks, the purgatory or intermediate state of Christianity which was called by medieval alchemists the astral plane."[34]

What Theosophists call the astral plane is a coarse plane of illusions. As we advance in meditation there may come a time where we begin to have hallucinations and visions, this is our consciousness entering the astral plane; or what I would consider a sub-plane of the pranic plane.

The astral sub-plane contains all of the apparatus necessary to create and maintain the illusion of reality in the physical plane. It is where maya is generated and the ability to sense things facilatated. Because it is the plane of illusions its inhabitants can have shape shifting ability and can be very deceptive.

Sri Aurobindo calls the plane of energy the vital plane and believed that it presented our current collective consciousness with its biggest challenges: "There are three obstacles that one has to overcome in the vital and they are very difficult to overcome, lust (sexual desire), wrath and rajasic ego. Rajasic ego(pride, stubbornness) is the supporting ground of the other two."[35]

Nature spirits, who oversee and maintain the plant and mineral worlds, co-habit the plane of energy in a separate subplane. As we shall later discuss many of the inhabitants of the pranic plane are focused on energy, its consumption and use and what throughout history have been called demons. The pranic plane is important because it gives life and sustenance to the physical plane, so if there is a problem that disrupts the flow of prana there will be problems such as storms and natural catastrophes in the physical plane.

The Mental Plane

The Mental plane is the plane of thoughts and mental machinations. It is where our thoughts find an independent existence once we have created them. A tremendous amount of competition goes on between thoughts in the mental plane. Since thoughts cannot manifest in the physical plane they look to influence us and try to get us to act on their behalf in the physical plane.

There is no need for speech in the mental plane because thoughts there can be read and understood the minute they are created. Everyone's thoughts are out there for everyone else to see and to know. Even if your consciousness or awareness is housed in the physical plane your thoughts can still be read by others in the mental plane. This means that what you are thinking at any given moment is available for all to know in the mental plane. This has broad ramifications for dealing with the inhabitants of other planes.

Thoughts from the mental plane can be reflective of greater thoughts, of our collective consciousness, or archetypes that can exert a tremendous influence upon us as Arthur E. Powell notes:

> Taking a still wider view of the mental plane, it may be described as that which reflects the Universal Mind in Nature, the plane which, in our little system, corresponds with the Great Mind in the Cosmos....The Universal Mind is that in which all archetypically exists; it is the source of beings, the fount of fashioning energies, the treasure-house in which are stored up all the archetypal forms which are brought forth and elaborated in lower kinds of matter during the evolution of the universe.[36]

As will be discussed later our collective (un)conscious exerts a tremendous influence upon us.

Higher Planes of Existence

There are higher planes of existence beyond the mental plane that are closer to spirit than the material worlds. Each plane gets us progressively closer to God and pure spirit. Each of the planes are associated with progressively higher states of consciousness,[37] as we become more loving, more willing to self sacrifice, have greater devotion and diminish our ego.

Several have a detailed cosmology for the planes above the mental plane.[38] I believe that it is difficult for us to truly know the higher planes of existence because we are biased by the influences of the physical and mental planes.

The purpose of planes of existence is to facilitate our spiritual evolution. As our consciousness evolves it assumes the attributes of the higher planes. (Note consciousness might not be the correct word to properly describe our spiritual evolution.)

Subtle Body and the Koshas

Just as the world around is made up of various planes so is the human body surrounded by invisible sheaths whose form resembles the human body but extends several feet beyond it. Hindu's call these sheaths that cover atman, or our true self, koshas. Others call them subtle bodies, or our aura. Christian paintings of Jesus, saints and other great souls often show the subtle body as a white glow, or arc halo, by the head of the character. Tibetan Buddhists believe that we have a gross (physical) body, a subtle (vajra) body and a very subtle body which is even more rarified. According to Vedanta the body has five subtle bodies, or koshas.

The koshas coincide with and correspond to the various planes of existence. Each kosha in our subtle body is simultaneously interacting with its respective plane of existence, although apparently quite unconsciously. In other words, the physical body is similar to and exists in the physical plane. The pranic body is similar to and exists in the pranic plane. The mental body is similar to and exists in the mental plane.

The psychic experiences, the out of body experiences, the premonitions, the telepathy and such that we sometimes experience has to do with our ability to transcend the physical body and place our consciousness in our subtle bodies and their respective planes of existence. It is that movement of consciousness we associate with our sixth sense.

The various koshas, or subtle bodies, are as follows:

- The annamaya kosha is the physical body. This is where we experience the physical senses and matter.
- The pranayama kosha, or energy sheath, is the next subtle body above the physical body.[39]

    The pranayama kosha contains the prana (life force) that powers our physical body and is vital to our health and well being. We absorb the prana to sustain us from the pranic plane. So if there are problems with the distribution of prana in the pranic plane it can affect our health. Prana enters our body during our mother's fifth month of

pregnancy.
- The manomaya kosha is the mental body that surrounds the pranayama, or energy kosha. It is where the mind resides.

  The manomaya interacts with the mental plane.
- The vijnanamaya kosha surrounds the manomaya kosha. It is the seat of discretion, knowing and intuition.
- The anandamaya kosha is "the sheath of bliss where pranashakti is united with the supreme self."[40]

The Subtle Body is Penetrable
Our subtle body is more viscous than it is rigid and interacts with all that it comes in contact with. C. W. Leadbeater says that our subtle body is penetrable and absorbs its surroundings wherever it goes.[41]

How permeable our subtle body is and how affected it is by the environment is difficult to say. The environment in this case is considered to be the unseen world and all that is swirling about within it. What is not in doubt is that part our aura/subtle body, rubs off wherever we go and some of the geographic locations' essence we visit stays with us. Our subtle body will similarly interact with the subtle body of people that we meet. This is an ongoing dynamic process that is constantly occurring.

The influence of the immediate environment upon our subtle body is very similar to how temperature affects our physical body. The overall consciousness of a particular location will act to influence our subtle body by either raising or lowering our individual consciousness, or vibe. The particular focus of a location—love, friendship or a violent act will also attempt to attach to our subtle body and influence us to do the same.

Our subtle body is also affected by more than our immediate environment. The thoughts and actions of others can influence and attach to our subtle body, as can entities, or inhabitants from other planes. The idea that our subtle body is penetrable and influenced by the environment underscores our interdependence.

There are no protective garments, no space suits, no sophisticated latex body suit that we can construct to prevent the exchange of matter with our subtle body. Our subtle body will be affected by where we go and with whom we interact. No matter how ephemeral or transitory the meeting may be, we will be affected. The influence will vary upon the strength of our subtle body and a variety of other factors. The most important determinant is the time spent with someone or someplace. The greater the time we spend with someone or someplace, the greater the influence.

Since our subtle body is influenced by our environment, then the character of the environment we live in, work in and travel through becomes very important. I am talking about the influences of the other planes of existence. Disturbances in the pranic plane can rob us of the precious life force of prana if Mother Earth's subtle body has been compromised. Thoughts in the mental plane can drive our thinking and actions to the point that we are the mercy of them. As we are about to find out in the next chapter the mental plane is very polluted with negative thoughts and consciousness. All the time our subtle body is being bombarded with polluted consciousness.

Chapter 2

Thoughts, Thought Forms, Samskaras

Our thoughts have a power much greater than most of us understand or can imagine. They can help or hurt us. They can bind us, chain us and make us slaves unto them; they can even have us worship them. They can cause us to create other thoughts, or have us take actions that build more bad karma that must be fulfilled, thereby leading to many more rebirths. They can also help liberate us and others. They can help us individually and collectively to evolve.

Sages and prophets since time immemorial have preached about the potential evils of the mind and the necessity to control it. Sikhism tells us to, "conquer your own mind, and conquer the world."[1] Hinduism and Buddhism advocate specific practices of meditation to control the mind and ultimately to bring about liberation and unite with the divine. The Bhagavad-Gita states that, "For him who has conquered the mind, the mind is the best of friends; but for one who has failed to do so, his very mind will be the greatest enemy."[2]

Unfortunately we have not learned to control our thoughts, so the mind remains our enemy. This failing has created a much larger collective problem that now threatens the world. Our violent and selfish thoughts and the consciousness associated with them hover around us like a bird of prey looking to manifest and influence us.

Thoughts Have an Existence, Power

Our thoughts are actual things. When we have a thought we create what some call a thought form. Each thought we have takes on a life of its own. It finds existence in the unseen world around us. Invisible to the naked eye, thoughts can be felt and seen by a sensitive person.

Thoughts have a consciousness associated with them and as the Bhagavad-Gita tells us, consciousness can run from the divine to the demonic.[3] So the thoughts around us similarly run the continuum from the divine to the demonic.

William Walker Atkinson a major influence early on in the New Thought Movement says our thoughts are as real as electricity or light, "When we think we send out vibrations of a fine ethereal substance, which are as real as the vibrations manifesting light, heat, electricity, magnetism."[4]

Affirmations

Anyone doubting the power of our thoughts need only consider the use affirmations to bring about dramatic changes in health, personality and help overcome fears. Affirmations are the act of affirming, repeating, or stating an assertion or statement continually. Positive affirmations such as 'I am healthy, I will succeed, I am happy' have been used successfully by self help gurus, psychologists, psychiatrists and others to bring about dramatic improvements in people riddled with fear or lacking confidence. It has been demonstrated that by simply repeating an affirmation, whether we believe it or not, will bring about results.

Transpersonal psychologist Frances Vaughan notes the power of affirmations:

Some of the benefits of spiritual affirmations include incrassating feelings of freedom and inner peace, an expanded capacity for compassion and sympathetic joy and a reduction of fear, anxiety and depression. Any quality can be consciously cultivated by consistent use of affirmations. Spiritual values suggest that the methods for cultivating qualities and training attention are best directed to goals that transcend ego.[5]

There is no difference between our thoughts and affirmations, all thoughts are affirmations and all affirmations are thoughts. The use of affirmations to overcome mental and emotional handicaps demonstrates the power of a thought to triumph. This is especially true when that thought is contrary to the practitioner's beliefs and ends up conquering their greatest fears.

Affirmation is very similar to mantra, the repetition of a phrase, word, or syllable. Some believe that repeating certain syllables creates a particular vibration that elevates consciousness. Others believe that it is the phrase itself that raises the consciousness by repeating and focusing on the same divine thought. Mantra is one of the easiest methods to use to enter the trance-like state of meditation. Repetition is powerful.

It is important to note that affirmations can create problems. If our thoughts are destructive or violent we are de facto reaffirming destruction and violence. Consider how many of us spend our leisure time. Some of us listen to music genres whose lyrics are violent, demeaning of women and reinforce racial and ethnic stereotypes. We play violent video games, watch violence on TV and watch violent sports. I am sure that we can all think of many other examples of how our thoughts and affirmations induced by our cultural milieu are reinforcing the downward spiral of our individual and collective consciousness.

Vaughan goes on to describe the debilitating consequences of negative affirmations:

When affirmations are used in the service of ego goals, such as attaining worldly possessions, they may have a backlash. 'Be careful what you want. You may get it' is a caution worth heeding. Getting what the ego wants can sometimes imprison the soul. Sometimes an unexpected side effect sparks fear and anger. The power of positive thinking is familiar to many people, and the shadow side of it, the unconscious backlash, has become familiar to psychotherapists. The consistent use of affirmations can be powerful, but any powerful tool can be abused. The misuse of affirmations in the service of ego defenses such as denial or repression can have troublesome repercussions.[6]

Unfortunately, the repetition of dubious claims and lies have been used all too often in business and politics to manipulate public opinion, whether it be to sell a product, or rush to war. Gustave Le Bon in *The Crowd* notes how politicians can sway public opinion: "When an affirmation has been sufficiently repeated and there is unanimity in this repetition...what is called a current of opinion is formed and the powerful mechanism of contagion intervenes."[7]

Thought Forms, What are They?

Annie Besant and C. W. Leadbeater in *Thought Forms* describe the visual appearance of thoughts as, "Each definite thought produces a double effect—a radiating vibration and a floating form."[8] The idea that thoughts have a form, a cloud like appearance, is why Besant and Leadbeater call them "thought forms". Thought forms are primarily made up of consciousness and should be considered to be reflective of consciousness.

Our thoughts are more than mental machinations and a reflection of our desires. They are all that we think and do. When we feel love, hurt, get angry, or are depressed, no matter what the emotion, we are creating thought forms. Similarly, our actions whether they are deliberate or not, physical or verbal, are thought forms. When we shake hands, kick a ball or brush our teeth we are creating thought forms. While it may seem a fantasy to say that our actions are in reality thought forms, we must remember that the physical plane is an illusion created by maya. We think that we are shaking hands, we think that we are kicking a ball, we think that we are brushing our teeth—the reality is that all we are doing is creating thought forms. These thought forms through the cloak of maya have us believe that we are doing things.

Because our motivations are often grey the composition of our thoughts is complicated. Our thoughts may be motivated by anger or a response to envy or we may have an ulterior motive in mind. I would view these different motivations and intentions as creating several thoughts.

For example, we might be thinking about how we can help someone. At the same time we might be thinking how helping someone benefits us. While they are the same thought there are two components, or thoughts to it. The first is a genuine desire to help and sacrifice, while the second is selfish. Most of our thoughts are similarly complex so we are creating several thoughts simultaneously. That is why it is so important to endeavor to always have positive thoughts and actions. Negativity no matter how justified it may seem creates a seed thought of negativity that will blemish us and look to spread. So if we get angry and have violent thoughts towards someone that has injured us, even if we don't act on it, we are creating a negative seed thought. We must always endeavor to turn the other cheek.

Generic Nature of Thoughts

Besant and Leadbeater note that there is a generic consciousness to each thought:

It should be understood that this radiating vibration conveys the character of the thought, but not its subject. If a Hindu sits rapt in devotion to Krishna, the waves of feeling which pour forth from him stimulate devotional feeling in all those who come under their influence, though in the case of the Muhammadan that devotion is to Allah, while for the Zoroas-trian it is to Ahuramazda, or for the Christian to Jesus. A man thinking keenly upon some high subject pours out from himself vibrations which tend to stir up thought at a similar level in others, but they in no way suggest to those others the special subject of his thought.[9]

The idea that our thoughts have a generic nature is a very important principle. While our thoughts may be focused on one person, place or event, each thought has a universal nature to it, love, hate, devotion, which transcends the specifics. The idea that a Hindu rapt in devotion to Krishna encourages similar devotion of Muslims towards Mohammed or Christians to Jesus applies to all of thoughts. If we are thinking lovingly towards someone we are simultaneously sending out love to others and encouraging them to think lovingly as well.

Similarly, if we are harboring bad thoughts towards someone we are at the same time encouraging others to have bad thoughts too. The character of our violent thinking can influence loved ones close by and strangers far away. Consider the hunter[10] who stalks its prey for hours and often days on end. All the time spent thinking about stalking and killing creates a very vile and predatory thought form. The generic character is the predatory behavior of stalking and killing. It is akin to the mindset of a predator like a rapist, child molester and murderer. It gives strength to this thought form and type of behavior in other others.

The generic nature of a thought is not confined to the character of the thought. A variety of generic factors are created with each thought. For example, if a thought is created upon impulse, then we are creating an impulsive thought, a thought whose nature is related to the process of being impulsive. If a thought is created because of something we saw, we are creating a response thought, or being reactionary. This can begin to develop reactionary behavior. We will also create associations with thoughts, where we have the thought, or what we were doing at the time.

The Moral Component of Our Thoughts

There is also a moral component to our thoughts. Jesus taught us that it is not necessary to carry out a sinful act to be a sinner, but to only think about sin:" *You have heard that it was said, 'You shall not commit adultery.'* But I say to you that everyone who looks at a woman with lust has already committed adultery with her in his heart."[11]

The moral component of our thoughts, whether they are loving or sinful, has a price, or karma associated with them. Loving and giving thoughts and focus on the divine will attract more of the same and elevate our being. Selfish and violent thoughts or thoughts of wealth and greed will diminish our being and create more karma that needs to be paid back. James Allen notes how our thoughts are a reflection of our character; "Thought and character are one, and as character can only manifest and discover itself through environment and circumstance, the outer conditions of a person's life will always found to be harmoniously related to his inner state."[12]

We Become What We Think About

Not only can affirmations, repeating or thinking the same thing over and over, get us to change our behavior but it will cause us to become what we think about. Those that have ever used the meditative technique of focused concentration on an object are familiar with how as our meditation progresses we get closer to the object being meditated on. Ultimately as Swami Satyananda Saraswati tells us, that at one point we merge with the object we are meditating on, "When you practice trataka[13] or meditation on an object like a shivalinga, the vrittis[14] diminish slowly, and ultimately there is a sudden flash of consciousness when the mind fuses completely with the object."[15] It is through this merger with the object of meditation that we learn its true nature.

It is not necessary to go into deep meditative states to merge with an object, whenever we begin to think about something or focus on it we begin the process of merging with it. That it is why it is so important to sincerely focus on God all the time so that we get closer to God and develop divine qualities.

Unfortunately so much of our time and effort is spent focusing on material objects or negative thoughts that we become those negative things. Swami Satyasangananda Saraswati in *Sri Vijnana Bhairava Tantra, The Ascent* her translation of Sloka 88 tells us that we become one with that which we meditate upon. In the commentary she goes on to say, "Whatever the self identifies with becomes the form of consciousness. This is known as self-identification. When you identify with the pleasures of the world, the mundane relationships, the material objects, that is the form that your consciousness assumes. But if the self identifies with transcendental awareness, the consciousness assumes the form of divinity."[16]

We are unknowingly bombarded daily with affirmations, from political propaganda to advertising slogans. Nutritionist's say that 'we are what we eat,' meaning our health and well being is a function of our diet. Similarly we become what we think about and focus on. This is a very important mystical truth for us individually and collectively, because whatever we apply our thought to we become, or, at minimum, take on some of its attributes. Again this is why it is so important to focus on the divine and the positive rather than the negativity and violence of the material world.

To Deny is to Affirm

Our negative thoughts can debilitate us. Frances Vaughan notes how negative affirmations end up reinforcing what we are trying to overcome:

"Affirmations are the most effective when they are short and succinct. They should always be stated in positive terms; for instance, 'I am calm and confident' is a positive statement, whereas 'I am not afraid' is a negative statement. The subconscious mind tends to be quite literal, and what is remembered is the word 'afraid', while 'not' is forgotten. Fritz Perls, the founder of Gestalt therapy said, 'To deny is to affirm'. For example, saying, 'I wouldn't think of leaving you', means I have already thought of it. Since the mind cannot deny anything without first thinking of it, affirmations should simply state what one wants to affirm."[17]

Similarly sages tell us that we should never be against anything, because once we are against something, we give it strength. This is a very important principle to keep in mind.

We are constantly giving strength to things through our resistance. For example, many activists, while ardently against war and willing to suffer immensely to stop war, are giving strength to war by protesting against it. There are many more examples that each of us has in our lives where we nobly try to stand against something or prevent something but in doing so give strength to and reinforce what we are trying to stop.

Science is beginning to realize that by thinking about the negative we give it strength:

Forget about the threat that mankind poses to the Earth: our very ability to study the heavens may have shortened the inferred lifetime of the cosmos...

New Scientist reports a worrying new variant as the cosmologists claim that astronomers may have provided evidence that the universe may ultimately decay by observing dark energy, a mysterious anti gravity force which is thought to be speeding up the expansion of the cosmos.[18]

At the heart of Tantric thinking is the realization that trying to stop or suppress something is not productive. Both Tantric Hinduism and Buddhism believe that we cannot stop our base behavior, but can overcome them through transformation. Tantrics believe that by putting forth and focusing on a positive alternative such as love or the divine we can liberate ourselves. One of the practices Tibetan Buddhists advocate is meditating on a deity and imaging yourself assuming those divine qualities.[19]

Whatever we are against we give strength to. Actions such as war, to protect or bring about peace, promote and reinforce violence. So much of our world is based upon being against something. For example, in the field of medicine the use of specific vaccines or drugs to eradicate a certain disease has led to the disease or microbes mutating and developing new and more virulent strains. While it seems preposterous to consider stop using certain drugs that are meant to kill or eradicate a disease, we need to realize that we may be creating an incubator for more hideous diseases for generations to come.

Whatever we apply thought to, whether in a positive or in negative way, we give strength to it through our thoughts. Thought is power.

Once Created Thought Forms Become Independent

Once a thought is created it is unleashed and looks to manifest and carry forward its design and intent. If the thought is projected at a specific person the thought form will try to attach to the recipient's subtle body and exert its influence. Whether the thought form is absorbed and the extent of its influence depends upon its intent, whether the recipient is open to the thought, the spiritual strength of the recipient's aura and whether there are similar thoughts attached to the recipient (similar thought attachments would increase the proclivity for success). A negative thought directed at an individual can bring about a 'return to sender' effect and reinforce the notion that you 'reap what you sow.'

A thought form will look to exert its influence and it will do so in a very headstrong and dedicated fashion. A variety of factors from its strength, to what it meets along its path and what or for whom it is intended will determine its success. The thought form may never reach its objective, or look to manifest its will on someone else. It is important to note that we do not have control of our thoughts once they are created. If our mind is full of violent and vile thoughts be aware that they can easily manifest in loved ones close to us. Once created our thoughts often meet up with other thought forms and form a contagion, of like minded thoughts, either for the good or for the bad.

Our thoughts and actions reverberate around the globe and can influence and are influenced by much larger issues. Thich Nhat Hanh tells us that, "The daily wars that occur within our thoughts and within our families have everything to do with the wars fought between people and nations throughout the world."[20] If we want love in the world we need to have love in our hearts and our minds.

Earlier it was noted that repeating musical lyrics of violence, hate, or racism brings that consciousness into us; it also unleashes a symphony of similarly like-minded negative thought forms upon the world. The thoughts associated with whatever you think, do, or watch are sent out into the world, 24/7.

The Law of Attraction
Like minded thoughts are attracted to each other. The idea that our thoughts could gather to form some sort of union or gang may seem a bit far fetched for some, but as Madam Blavatsky notes, mystics and adepts have long known that like attracts in the unseen world: "It is the universal law, which is understood by Plato and explained in Timaeus as the attraction of lesser bodies to larger ones, and of similar bodies to similar."[21]

William Walker Atkinson said that there was one great law in the universe, the Law of Attraction:

We speak learnedly of the Law of Gravitation—but ignore that equally wonderful manifestation—The Law of Attraction in the Thought World. We are familiar with that wonderful manifestation of Law which draws and holds together the atoms of which matter is composed. We recognize the power of the law that attracts bodies to the earth, that holds the circling worlds in their places—but we close our eyes to the mighty Law that draws to us the things we desire or fear—that makes or mars our lives.[22]

The law of attraction says that whatever we think about or do will attract more of the same into our lives. This has the potential to create a strong pattern whereby we are continually besieged by the same type of thought or behavior.

Competing Thoughts

The mental plane is full of thoughts all vying to exert their intent. Sri Aurobindo's spiritual partner the Mother, Mirra Alfassa, describes the life of a thought once created, the battles, the desire to manifest and how filled-up the unseen world is with our thought forms:

We are all the time, always, creating images, and creating forms. We send them into the atmosphere without even knowing that we are doing so – they go roaming about, pass from one person to another, meet companions, sometimes join together and get on happily, sometimes create conflicts, and there are battles... If our eyes were open to the vision of all these forms in the atmosphere, we would see very amazing things: battlefields, waves, onsets, retreats of a crowd of small mental entities which are constantly thrown out into the air and always try to realise themselves. All these formations have a common tendency to want to materialise and realise themselves physically, and as they are countless – they are far too many for there to be room enough on earth to manifest."[23]

What goes on in the mental plane influences our thinking. If the mental plane is filled to the brim with thoughts and groups of thoughts vying to manifest then our minds are bearing the brunt.

Strong Thoughts

It is the interplay of mental activity, emotion and action that determine the vitality of a thought. If they are in sync and have the same focus, it will be a strong thought form. For example, a mental idea associated with an emotion is stronger than the mental idea by itself. So if we are praying (thinking) for someone that is sick and are emotionally wrought, the prayer will be stronger because it has an emotional component attached to it. If there is a divergence in focus then the thought will be weaker and easily dissipated.

Other factors that determine the strength of a thought are the spiritual and concentrative ability of the person creating the thought. The more spiritually evolved someone is and the greater their concentrative ability, the stronger the thought will be. The concentration and effort applied to a particular thought will also determine its strength.

Action if in harmony with a thought will give it added strength.

Samskaras

While each of our thoughts is sent into the world once it is formed, some of its essence remains with us. Hinduism teaches us that each thought (thought, action, emotion) we have creates an impression, called a samskara, in our mind. Once created our thought impressions, or samskaras, look to reassert themselves by causing us to repeat the same thought or do the same thing. It is said that each impression or thought we have is like a seed that is capable of bearing fruit (producing more like minded thoughts) and producing many more seeds. If we think or do the same thing over and over again the deeper that impression becomes, the more it begins to exert an influence upon us. Over time if samskaras get deeper, they begin to cloud our thinking and exert an influence over us; if allowed to grow they can begin to dominate and control our lives.

Who has not fallen victim to a thought? We think one thing and the next thing we know we cannot get it out of our mind. We obsess about it, fret and fall victim to its sway. Samskaras if left un-checked can create obsessive, compulsive and addictive behavior.

Not only do our thought impressions, or samskaras, plant more seeds that bear fruit within us, they attract other like-minded thoughts. They are like a magnet attracting similar thoughts to us, thereby putting more pressure upon us to repeat the thought, or action even more. If we put out love, then it is love that we will attract. If we put out violence, then violence will be drawn to us. So it is critical that we thinking positively lest we cast seeds of negativity that will spread with the vigor of weeds.

Samskaras are a permanent record of all of our thoughts and actions that stays with us. If we love someone it will be duly noted; similarly if we hurt, or even kill someone it will also be recorded. It is the generic nature of the thought or action that is permanently recorded. Once attached to us samskaras stay with us for a long time, particularly if the impression was deep. Samskaras are subject to the law of karma, and are removed when we reap what we have sown.[24]

The Root of Karma

If samskaras are not cleared away before we die we carry them forward as karma in our future lives. Samskaras are the root of our karma as Swami Satyananda Saraswati of the Bihar School of Yoga states in his analysis of Patanjali's Yoga Sutras: "It is said by Nagarjuna that a seed when not burned is capable of giving rise to many seeds and plants. In the same way when the chitta (individual consciousness, memory) is not freed from the samskaras, it is capable of producing many more samskaras, bodies and reincarnations."[25]

Swami Vivekananda describes how samskaras cling to us through many rebirths:

> Each work we do, each thought we think, produces an impression called in Sanskrit samskara, upon the mind; and the sum total of these impressions becomes the tremendous force which is called character. The character of a man is what he has created for himself; it is the result of the mental and physical actions that he has done in his life. The sum total of the samskaras is the force which gives a man the next direction after death. A man dies, the body falls away and goes back to the elements; but the samskaras remain, adhering to the mind.[26]

Many enlightened persons say that samskaras are the greatest impediment to our spiritual progression because they are the source of bondage and keep us chained to the cycle of birth and death.[27] Swami Vivekananda calls the endless cycle of death as, "samsara in Sanskrit, literally the round of birth and death."[28] Similarly Buddhists, Jains and Sikhs refer to the cycle of reincarnation as 'samsara'.

At some point we have to remove our samskaras. Samskaras are synonymous with thought forms. Samskaras are thought forms and thought forms are samskaras. The fact that they can transcend the cycle of life and death shows how powerful they are.

Mental Baggage

We are surrounded by our thoughts and actions, both past and present. We always carry with us the record, or luggage of sorts, of all that we have done and thought. Some refer to this as our vibe, many can sense it and some can even see all the thoughts circulating around us. Mirra Alfassa gives a description of the appearance and behavior of certain thought forms as told to her by a friend that could see them: "[H]e had always noticed that people who have sexual desires are surrounded by a kind of small swarm of entities who are somewhat viscous and rather ugly and which torment them constantly, awakening desire in them.... It was like being surrounded by a swarm of mosquitoes, yes!"[29]

Whatever we experience or think about we carry about with us and attract the same. If we experience something bad, an accident or a tragedy, a piece of it will be attached to us. This can create a propensity for similar accidents or tragedies over the short term, particularly if they were emotional events. It could also be that someone close to us has been traumatized by an accident and we absorb some of their samskaras that induced an accident. The saying 'when it rains it pours.' or that 'things come in two's or three's,' means that if something bad happens it often triggers similar events, a sort of chain reaction.[30] We have also seen the same repetitive behavior with more pleasurable events where grand prize winners win again and again.

Geographic Samskaras

Not only do our thoughts project into the world and attach to us as samskaras, they remain where they were created. Wherever we have a thought (thought, action, emotion), an impression is left at the location creating what I call a geographic samskara. Their strength varies, a very strong spiritual experience or a violent emotional experience can create a powerful geographic samskara. Geographic samskaras like thought forms are primarily made up of consciousness.

We also leave an energy trail from our thoughts and subtle body. Later in Chapter 12 "Sacred Earth—Creating Sacred Space", we will talk about how this energy trail can be used to create certain structures to enhance our spiritual experience.

Renowned psychoanalyst Carl Jung believed that land could influence people. Specifically he felt that the consciousness of indigenous people permeated their land and would influence foreigners that came to live there:

> Certain Australian primitives assert that one cannot conquer foreign soil, because in it there dwell strange ancestor-spirits who reincarnate themselves in the new born. There is a great psychological truth to this. The foreign land assimilates its conqueror.... Everywhere the virgin earth cause at least the unconscious of the conqueror to sink to the level of its indigenous inhabitants....Our contact with the unconscious chains us to the earth and makes it hard for use to move, and this is certainly no advantage when it comes to progressiveness and all the other desirable motions of the mind. Nevertheless I would not speak ill of our relation to good Mother Earth.[31]

Like all thought forms geographic samskaras try to manifest themselves and multiply; only they do not move and remain in a particular location. The world is blanketed with geographic samskaras and we are constantly in contact with them. If it is a place with divine consciousness, a place where people pray, then that divinity will attach to us to some degree. If it is a place where violence has happened, then we will be struck by violent thoughts and possibly even have violence come to us.

Often we see the same behavior occurring over and over again at the same location. Notice how animals are hit by cars in the same location over and over again. This repetition of action and thought adds to the strength of a geographic samskara.

We should not make assumptions as to what is a good place or a bad place. It is only our sixth sense and ability to sense consciousness that can tell us about a place. For example, it would be wrong to assume that a place of worship is going to have a divine consciousness. While praying or having- elevated thoughts raises consciousness at a place of worship, the participants can negate those beneficial affects. People bring their own baggage, negative samskaras—violent thoughts, greed, hate, etc., when they come to worship. Some of those samskaras attach to a place. Like every place else houses of worship are grey, containing both good and bad, hopefully more divine.

Sacred Space

Wherever we go we encounter geographic samskaras and thought forms. Some of that consciousness attaches to our subtle body and we leave some of ourselves behind. Geographic samskaras act to influence our consciousness and can either raise or lower it like temperature changes. We are constantly being influenced by the consciousness of places where we are, either for the good or bad. Like individual samskaras, these geographic samskaras can be a huge impediment to our spiritual growth. For example, we can be a very devout person, compassionate to others and follow a very dedicated spiritual regime but all that can be significantly diminished by negative geographic samskaras.

Swami Satyasangananda Saraswati notes how our environment can influence our spiritual development:

Along with commitment to the practices, another important expedient for overcoming obstruction in sadhana is to live amongst people who have a raised awareness. This helps a great deal as their heightened awareness uplifts all those who come in contact with them, resolving many obstacles effortlessly. This contact is possible in an ashram where the environment is built up over the years by the positive vibrations of the people who inhabit it, so that even a chance visits to such a place is very uplifting and resolves many difficulties and obstructions in one's sadhana.[32]

Unfortunately most of us do not live in ashrams or very holy places where evolved souls and positive geographic samskaras focused on the divine reside. The fact is that the majority of the world is covered with negative geographic samskaras that have been building for millennia. These continually diminish and influence our consciousness, mostly for the worse. These negative samskaras pose a major threat to the world and are a huge impediment to our collective spiritual advancement.

Talismans and Amulets

We dynamically interact with the objects that we come in contact with, or close proximity to. Our thoughts and consciousness attach to objects that we encounter. At the same time our subtle body absorbs some of the consciousness of what we come in contact with. The longer we carry an object, the more that it retains part of our consciousness and the more we absorb its consciousness. I have seen clairvoyants that have been able to gain insights about people simply by holding an object of theirs. Some dowsers that I know refuse to lend their dowsing rods for others to use because they do not want to have them pick up the other person's consciousness fearing it will bias their readings.

Objects can be intentionally charged with our consciousness, or for particular purposes. Occultists have long understood the power of good luck charms, amulets and talismans to protect and influence people. A talisman is any object that has been charged with the intention of its maker to create change.[33] It can be a stone, a doll, a remote control or any other object. Often a ceremony or ritual is performed to increase its strength. As with thought forms the power of a talisman is determined by the intention, focus and emotion applied to it.

Not only does our consciousness attach to an object but we are constantly creating talismans, whether intentionally or not. Each object we create is a talisman with a consciousness associated with it. Our homes, the TV, the computer, the car, the bench, our coat, etc., are all talismans. Each object radiates a consciousness that we absorb and our world is full of them. Don't you find it odd that the materialist and capitalistic system advocates the ownership of things (talismans)? Why? Isaiah and the other Hebrew prophets railed against idols because of their power to influence:"Their land is filled with idols; they bow down to the work of their hands, to what their own fingers have made."[34]

Is not the TV and computer a tribute to technology, or the car a symbol of our triumph over the ability of our legs to walk?

There are also talismans of violence such as guns and other weapons whose only purpose is to kill and hurt. Weapons carry a particularly vile thought form of killing and violence and should be avoided if at all possible. Care should be exercised in your contact with people carrying weapons for whatever purpose.

All of these objects are dead things, although with genetic engineering science is trying to change things. They have no life, they are artificial things. They exude death. Remember the words of Jesus, "For where your treasure is, there will your heart be also."[35]

Violent Talismans

Whatever we apply thought to can become a talisman and over time it can become quite powerful. The desire to conquer and kill has culminated in a thought form that has given us the ability to wipe out all of humanity. For eons upon eons humanity has thought about how to kill and conquer and put thoughts into action and in the process created an infinite number of seed thoughts that have all borne fruit. From those earliest days there was the thought of violence done with bare hands and then primitive weapons of clubs and sharp instruments. Then there was knives and swords and eventually guns. The atomic bomb was an inevitably given all the intense focus on violence. There are many other revolting things in the world that we have created over time by continually thinking about the negative and violence.

Prayer

One of the most powerful, constructive things we can do is to pray. The power of prayer to bring about positive change is tremendous. Prayers can be intercessory if we ask for help, they can be lamentations asking to hear God, they can be adoration, a confession, an invocation, giving thanks, visualization; arguably, each thought we have can be a prayer. I would define prayer as a thought made for a greater purpose or connection to the divine.

I often interchange the word prayer with spiritual exercises such as meditation, contemplation and pranayama. Although I would also consider a prayer to be a specific request, plea or purpose.

Prayer is a targeted thought form that we create and goes out into the mental plane with other thought forms and seeks to manifest itself and do its work. If done with the best of intentions and in the emotional spirit of God, it can be powerfully transformative. The research and analysis of Dr. Larry Dossey has confirmed the healing power of prayer.[36]

When we pray for someone or something we should always try to give up control and not make decisions about what is best. Our prayers, or thought forms, will know best, as will God, on how to help someone. It may well be that the best help for someone is something other than we imagine. Remember that thoughts know best where to manifest themselves. Give up control.

Arguably each positive thought that we have is a prayer and has the ability to do good and help others.

Thoughts are Power

The idea that thoughts seek to produce more thoughts, influence us even to the point of control and are constantly trying to manifest shows how powerful they are. Our history of war and conquest show that our thoughts have a long record of being negative and focused on sense objects; consequently we have turned the world into a dystopia of sorts. Part II "Hell; Our World, Our Making" will discuss in greater detail how we made this bizarre mess and how we have let ourselves become slaves to our thought forms from science, to capitalism, to corporate power. Once a thought seed has been planted it takes root and bears fruit. If many people think about the same thing for many years it gets to be very, very strong.

Always remember that you give strength to whatever you think of. Choose wisely.

Chapter 3

Sweet Mother Earth

Mother Earth is a living being who nurtures us and gives us life. She blesses us with the food we eat, the water we drink and the air we breathe. The benefits she bestows upon us go well beyond physical sustenance and encompasses our spiritual well being and evolution.

Mother earth connects us. She links together, big and small, rich and poor, community and country, country and world. She facilitates connection to the divine. Through her veins flow the divine qualities of love, compassion, ahimsa, mercy, self-realization, humility, devotion and divine knowledge.

We cannot fathom all that Mother Earth provides. She is our mother.

We dynamically interact with Mother Earth in a symbiotic relationship; our thoughts and actions affect her. Like us, Mother Earth has a mind body relationship, only it is our minds affecting her body. We can make her ill through bad living, violence, and disregard for each other and for the environment. She responds to earth illness with violent storms and earth disturbances in the physical plane. When we make her ill we are deprived of some of the benefits she is meant to bestow upon us.

She longs for love, compassion and devotion to the divine and responds accordingly. Her body gains strength through our right living, our right mind and love for each other. When uplifted by our loving thoughts and actions, Mother Earth creates a more nurturing and spiritually fortifying environment for us. Our spiritual evolution is dependent upon Mother Earth.

Not only are we dependent upon Mother Earth, but her evolution is dependent upon us. Sri Aurobindo says, "There is a separate global consciousness of the earth (as of other worlds) which evolves with the evolution of life on the planet."[1]

Matricide

Geoffrey Hodson believes that because we have failed to properly approach Mother Earth we do not understand her: "The approach to Nature by modern man is almost exclusively through action and his outer senses. Too few among her human devotees approach her in stillness, with outer sense quieted and inner sense aroused. Few, therefore, discover the Goddess herself behind her earthly veil."[2]

Not only have we not approached Mother Earth in stillness, we have raped and plundered her. Long ago many thought seeds were planted of seeing Mother Earth as the enemy to be conquered. For eons these thought seeds have borne a succession of fruit that have planted more seeds of abuse and conquest. We have clear cut, polluted and killed each other as well as desecrated Mother Earth without regard. The industrial revolution accelerated this process through increased mechanization that gave humankind greater and greater leverage for exploitation. The results reach well beyond pollution, destruction of the biosphere, extinction of species and global warming.

Scientist James Lovelock says that humanity is like a disease that threatens Mother Earth: "Humans on the Earth behave in some ways like a pathogenic microorganism, or like the cells of a tumor or neoplasm. We have grown in numbers and in disturbance to Gaia, to the point where our presence is perceptibly disabling, like a disease."[3]

Because our mind set has been to conquer we have planted few seed thoughts of understanding, love and how to nurture Mother Earth. Those that did love Mother Earth like indigenous people and pagans were slaughtered or converted and so their thought forms of love for Mother Earth diminished over time.

Consequently we have little love for her in our collective consciousness, or understanding of her.

We know little of Mother Earth beyond what our physical senses see. Our focus on plunder has taught us how to exploit her to extract natural resources, what to put into the soil to maximize crop yields, how to use her rivers to create power. This is misuse of our precious home, earth.

Krishnamurti tell us that "if you hurt nature you are hurting yourself."[4] We need to remember that all is Brahman and Mother Earth is part of God as well. She too has consciousness, a soul and much, much more.

Mother Earth's Hidden Benefits

Thanks to a few, such as Indigenous peoples, geomancers, feng shui practitioners and pagans, we do know a little about Mother Earth beyond what our physical senses can comprehend. Mother Earth is an intricate and sophisticated being with whom we dynamically interact and who exerts tremendous influence upon us.

Mother Earth gives us hidden benefits such as Schumann waves, which are beneficial electromagnetic waves whose movement resembles our own brain patterns. It is believed that Schumann waves help regulate our internal clock by affecting our hormonal secretions. NASA (National Aeronautics and Space Administration) had to install artificial devices to replicate Schumann waves for its space craft when it discovered that astronauts returning from space felt distressed and disoriented.

Construction materials such as concrete and metal can impede the movement of Schumann waves to reach people in structures made of such materials. For example, it is felt that the metal fuselage of a jet airplane blocks access to Schumann waves and thereby contributes to jet lag.[5] Our world is filled with devices that emanate electromagnetic waves from cell phones to electrical appliances that interfere with the smooth flow of Schumann waves.

The pranic plane, the plane of energy, distributes prana, which is the life force that sustains the physical plane and Mother Earth. It also sustains and gives us life; without prana we could not exist in the physical plane. There are varying degrees of prana, from the macro-cosmic down to the micro level that powers specific functions in Mother Earth's subtle body.

Cosmic prana is the highest form of prana and is more consciousness than it is energy. It can only be absorbed by humankind. We get it through right thought and action and through spiritual exercises such as prayer, contemplation and meditation. We do not absorb cosmic prana through energy exercises such as pranayama, energy work, light work, etc. Cosmic prana and other forms of consciousness in the unseen world when absorbed raise our consciousness making us more loving, giving, compassionate and devout. They are integral to our individual and collective spiritual development.

Earth prana, chi, is the life force that sustains us and the plant and animal kingdoms and consists more of energy than consciousness. It is what most people associate with prana, or chi or qi in feng should I call it earth prana because it is recycled through the earth and feeds the plant kingdom and helps maintain the earth closest to the surface. It is the highest form of prana that is involved with the physical earth. There are coarser forms of prana that sustain the mineral kingdom and the earth's architecture.

When we are deprived of Mother Earth's nutrients our health and well being suffers.

Energy Flow

The matrix regulating the flow of prana is similar to a home heating system where the hot air ducts carry and release the hot air into rooms and the cold air ducts re-circulate the air. Prana from above is distributed by earth ducts into the air that we breathe and later recycled back up. There are variations on this theme depending upon the type of prana. For example, earth prana is sent into the earth before it is recycled into the heavens. Prana is constantly in motion, and when it is not, it stagnates and loses its potency.

Prana descends from above into an invisible duct on the surface of the earth where it is distributed in all directions(360). Unless its course of direction is blocked, prana will move in a straight line until it gets in close proximity of an earth chakra. Like a chakra in the human subtle body, an earth chakra spins in a clockwise motion. This circular movement creates a vortex that draws the prana into the earth chakra. The type of prana will determine where it goes next.

Earth prana is recycled through the earth before it ascends back into the ether. It travels in the earth through nadis, invisible wire-like structures what some call energy lines. They are like nerves or veins; some are big or major, while others are tiny. They are similar to what acupuncturists would call meridians. They make up and support Mother Earth's physical body, the earth. A nadi eventually connects with a duct where the prana is sent into the ether. It is estimated that there are hundreds of thousands of nadis in our body and multitude of that in Mother Earth's subtle body.

Earth Grid

A system of energy lines similar to nadis forms an intricate series of grids within Mother Earth. The grids act like the skeletal structure of a large building.

Some of the grids underlying Mother Earth's architecture that power her have been classified. Hartmann Lines, or net, is a series of energy lines running north south and east west creating a grid. They were discovered by Dr. Ernst Hartmann. Curry Lines create a global grid of energy lines that go northeast to southwest and southeast to northwest. They were first discovered by Dr.

Manfred Curry and Dr. Wittmann.

Dowsers and geomancer's are often concerned about the intersection of various energy and water lines and the ill affect that they can produce. They have found that if you sleep over the intersection of several energy lines that it can contribute to cancer. Should you ever get cancer you should first investigate whether something as simple as changing the location of your bed might improve your health.

Linking Communities and Countries

Our collective and individual consciousnesses mingle with Mother Earth's consciousness in an ongoing and dynamic process. Consciousness, like prana, is carried by the ether of the air via thought forms and by invisible lines, or veins of consciousness. The lines that carry consciousness are called lines of consciousness (L.O.C.'s) or spirit lines. Some refer to them as 'ley lines.'

Spirit lines emanate from what I call Mother Earth's soul[7] in upstate NY. They are designed to carry God's love and highest aspirations for humanity around the world. They are thousands of miles long and link communities and countries around the world. They are generally 8 to 20 feet wide and can[8] exert influence well beyond that.

LOC's dynamically interact with the consciousness it meets along its path. If a spirit line passes through a place where there is great love then it will be influenced by love. Conversely if it passes through a place of anger or violence, it will be tinged by the violence. The key point is to realize that a spirit line has the ability to carry the consciousness from a very good thought/action, or very bad thought/action long distances. In other words they link communities, so what goes on inside a gated community will interact with what goes on within an inner city ghetto. They truly underscore the notion that we are each other's brother's and sister's keeper.

Since spirit lines are meant to carry Gaia's soul around the world they are a magnet for the spiritual. This is natural because we are longing to connect with the divine and send out a positive vibe into our community. Geomancers and dowsers have long noted that sacred sites are often linked by spirit lines. Alfred Watkins who is credited with discovering ley lines in England early in the twentieth century learned about them when he noticed how holy sites separated long distances lined up in a straight line. Looking for spirit lines is one of the primary things I look for when investigating sacred places for Mother Earth Prayers.[9]

Spirit lines connect and attract the spiritual and all forms of consciousness from the divine to the demonic. Spirit lines can link certain types of businesses (used cars, medical, ad agencies, etc., libraries, types of individuals, or violent acts.) For example in Syracuse, NY there is a spirit line I call the Prophetic Spirit[10] because it links many people who have a strong orientation towards social justice. This includes churches and individuals that are separated by miles in distance.

Intersections of spirit lines tend to be power points that draw people. It appears that the intersection of several spirit lines creates a seat of consciousness that is greater than the sum of its parts. They will often attract places of worship, or be a meeting place, or the home of someone spiritually motivated. There is a long history of this. Intersection points of spirit lines are conducive to prayer and meditation. In particular they are places you should go to for communication and connection to the divine. If you are looking for direction or answers, then go to an intersection of spirit lines. It is at intersection points that Spirit lines perform a variety of purposes beyond carrying consciousness.

How Mother Earth Responds to Disease
Bad consciousness, violent actions and thoughts, hate, negativity,

ultimately lead to disease. Swami Niranjanananda Saraswati of the Bihar School tells us that what is going on within our mind and heart eventually filters down to the physical body:

> The body is part of the consciousness manifesting externally. It is a unit of the cosmic, universal, all-pervading consciousness. The body does not contain consciousness, rather it is consciousness which expresses itself through the body at these different levels. Therefore, I would say that the body is never ill and never healthy. The body is simply responding to what is filtering down from this higher level, from the consciousness. If something coming down into the head centre is distorted, we go through head trips and have headaches; if what is coming down into the middle centers is distorted, we go through heartaches; and, if what is coming down from the higher to the lower centers is distorted, we go through a lot of frustration, aggression, suppression, anxiety, fear and insecurity. It is these mental states which later alter and influence the performances of the body and manifest in the form of illness and disease.[11]

Mother Earth, like us, is affected by bad consciousness. Only the bad consciousness comes from us and not Mother Earth herself. Our bad thoughts, violent behavior, exploitative and abusive motives and polluting mentality do great harm to Mother Earth and cause her to break down. Some still doubt that global pollution is creating global warming. Many laugh at the idea that our thoughts or violent behavior hurts Mother Earth. God tells us in Genesis that it was wicked behavior and evil thoughts that led to the great flood during the life of Noah:

> The LORD saw that the wickedness of humankind was great in the earth, and that every inclination of the thoughts of their hearts was only evil continually. And the LORD was sorry that

he had made humankind on the earth, and it grieved him to his heart. So the LORD said, 'I will blot out from the earth the human beings I have created—people together with animals and creeping things and birds of the air, for I am sorry that I have made them.' But Noah found favor in the sight of the LORD.[12]

Similarly Jeremiah said that Israel defiled the land by worshipping false gods and idols, "she polluted the land, committing adultery with stone and tree."[13]

Besides the great flood, the Old Testament is filled with story after story about how wicked behavior and bad thoughts created natural catastrophes. In Psalm 1.4 we are told that the wind will drive away the wicked.[14] Ezekiel 5.6-17 educates that famine is the consequence of living outside of God's law. Jeremiah 14.1,15,16,18 tells us that the consequences of living out of balance with God's teachings are famine, drought, and pestilence and that we even cause imbalances in the animal kingdom. Later Jeremiah adds that faithlessness to God and violence makes the earth mourn and parches the land: "For the land is full of adulterers; because of the curse the land mourns, and the pastures of the wilderness are dried up. Their course has been evil, and their might is not right."[15]

The global warming, pollution and litter covering the world and threatening our existence is symptomatic of a larger problem, our bad consciousness.

Blockages

While prana is the life force that sustains and gives us life, it is our collective consciousness that will determine how efficiently it is distributed across Mother Earth. As it is with our own self, negativity and demonic thoughts and behavior will create blockages in the proper flow and distribution of prana, and so it is for Mother Earth as well.

The physical manifestation of disease is "caused by improper distribution of prana in the physical body."[16] In other words, whatever part of our body is sick, it is so because it is not getting enough prana, and somewhere there is a blockage or broken flow. The blockages are caused by bad consciousness that disrupts the smooth flow of prana within our body. Acupuncturists and pranic healers all work with prana/chi to remove blockages in the nadis/meridians, redirect it and restore the pranayama kosha to its normal functioning.

Consciousness dynamically interacts with energy (prana) in our own and Mother Earth's subtle body in an ongoing process. Since energy devolves from consciousness it is consciousness that will determine how energy is influenced. The waves of thoughts in our mind are called vrittis, literally translated as "whirlpools"[17] because of their circular motion. Geographic samskaras are the equivalent of vrittis for Mother Earth.

Earlier we had noted that geographic samskaras display either a clockwise (positive consciousness) or counter-clockwise (negative consciousness) movement. As negative geographic samskaras get stronger they begin to disrupt the flow of prana. Initially they will cause earth prana to stagnant in a particular place, ceasing its movement, or changing its course of direction. This disruption, or blockage, means that the prana is not being properly distributed. Although we do not know for certain, it appears that these blockages may reduce the amount of prana available for us to absorb and thereby contribute to disease.

Over time as a negative geographic samskara gets stronger it begins to drive the earth prana into following its negative counter-clockwise movement, creating what dowsers call a negative energy vortex. They are basically mini-tornadoes that are very debilitating and rob you of energy and diminish your consciousness. As a negative energy vortex grows in strength it will begin to disrupt other higher forms of prana that contain more consciousness as well as other beneficial essences that

Mother Earth nurtures us with. You should avoid being close to negative vortexes.

The blockages of prana created by foul consciousness means that some part of Mother Earth is being deprived of the life force. Just as with our own body this disruption in the proper flow of prana leads to illness and disease for Mother Earth. The floods, droughts, famines, hurricanes, earthquakes and violent storms in the physical plane are all somehow tied to blockages in the flow of prana. Mother Earth is not only breaking down, but in the process, trying to cleanse herself of the foul consciousness that permeates her. Science tells us that the Coriolis Effect, the influence of the earth's rotation on objects, contributes to turning a violent thunderstorm into a hurricane or tornado by giving it circular movement (counterclockwise motion in the northern hemisphere, clockwise in the southern.) Not surprisingly, hurricanes and negative energy vortices have the counter-clockwise movement. Arguably a negative energy vortex is a mini tornado. Given this we must ask ourselves does foul consciousness in the form of negative geographic samskaras influence the making of a hurricane or tornado? Do they create a permanent dead spot for the flow of prana on Mother Earth's subtle body? More needs to be studied and learned, and to do this we need to have a perspective that we are living in a mystical world.

We have covered the world over with bad intentions. The atmosphere is laden with large thought forms of violence, greed and selfishness. The world is covered with massive geographic samskaras where war, genocide, torture, slavery, abuse and other forms of violence have transpired. The world is filled with talismans we have created, pollutants, industrial waste, electronic devices and appliances, motor vehicles and much more. Then there is looting and violation of mother earth through mining, excavation, drilling and deforestation.

Damaging Mother Earth's Subtle Body

All of these actions tear at Mother Earth and make her sick. They also add to and give strength to maya and hamper our ability to connect with Mother Earth, reducing many of her beneficial affects. They affect our consciousness and make us sick. Negative geographic samskaras beckon to us like a drug addict in a downward death spiral to continue the same bad behavior. Were Mother Earth not covered with negative geographic samskaras and talismans that impede the distribution of her beneficial essences our state of health, mentality and spirituality would be significantly better. Native American, Sun Bear, understood the consequences of disrespecting creator's message and abusing Mother Earth. Along with Wabun Wind in *Black Dawn Bright Day* they write:

> Native prophecies from all lands, and the spiritual teachers of many different peoples, speak of things that will happen at this time. They speak of major changes. They speak of people who will survive, human beings who will want to take a sacred path in harmony with the earth. They say that those who do so will stay alive even though there will come a time of great destruction to the Earth. The wise people will know what to do and will move in a sacred manner to make the changes necessary for their own survival and for the survival of others. Those who do survive will be the people who have studied the prophecies and have learned how to hear the Earth.[18]

The larger concern is that by damaging Mother Earth's subtle body we may be reducing her ability to properly function. This limits her ability to maintain the environment, potentially having detrimental consequences. For example, medical doctors are all too familiar with how a patient that is comprised from a serous illness is very vulnerable and can succumb to a host of diseases such as pneumonia, the flu or the common cold, that for the healthy person would mean bed rest but not death. In other words, it may well be that while pollution taxes Mother Earth, she has the ability to process a lot but if her subtle body is compromised her ability to deal with manmade pollution is significantly reduced.

Could it be that we have global warming because we have damaged Mother Earth's subtle body through our vile thoughts and violent behavior. Just as with our own subtle body, we have created blockages in Mother Earth's subtle body that are now beginning to manifest in the disease of global warming. No doubt we have damaged Mother Earth's subtle body and diminished her ability to properly function. How much of a role this is playing in global warming is difficult to determine at this time. What is not in doubt is that it is humanity's base consciousness is the root cause of our problems today.

Mothered Earth Responds to Love

As much as Mother Earth responds adversely to violence, hate and wicked thoughts, she responds favorably to love. According to Tantric Hinduism, prayer, meditation and other forms of spiritual exercises increases our level of energy. It holds that prayer/meditation helps raise the Kundalini[19] energy (coiled serpent) trapped in our mooladhara chakra (base chakra) that like cosmic prana is more consciousness than energy. Proponents of laya yoga, such as Shyman Sundar Goswami, feel that as long as our kundalini energy is coiled and dormant our spirituality is hidden and our focus remains on the physical world and sense objects. When she (kundalini energy) is aroused spiritual power and consciousness are manifest.[20]

As we spiritually evolve our kundalini energy rises piercing our various chakras. When it reaches our anahata chakra (heart chakra) it unites with our individual consciousness in what has been described as an explosion of bliss. United, or re-united, they together move higher and ultimately unite with global consciousness (God) in our Sahasrara chakra (crown chakra).

The union of energy with consciousness has significance in Hindu philosophy. It holds that the world began as pure consciousness/Godhead (Purusha), and from this the material world energy (Prakriti) was created. Consciousness is also referred to as Shiva and energy/matter as Shakti. The union of Shiva (male aspects of God) with Shakti (female aspects of God) marks the beginning of the sojourn back to God and the reversal of devolution into the material world.

Swami Satyadharma says that the rising of kundalini begins our inner experience and lifting the veil of maya:

At a certain point in man's evolution, kundalini begins to awaken and arouses man from the dream. With the ascent of kundalini, man begins to experience the inner reality. The veil of maya is gradually removed as the boundaries of individuality are dissolved. With the rising of kundalini, one's self identification progressively expands to include the universe. Thus it is said to be the same kundalini which creates and binds the jiva in the body, and which also withdraws the bonds by revealing the process of liberation."[21]

The same process of kundalini energy rising within our individual body is also at work within Mother Earth. When we pray, meditate, have positive intentions or do good loving things we can attract energy to a particular location. Dowsers such as Sig Lonegren have long noted that energy lines, undetectable to the human eye, are drawn to sacred sites. They have found that over time continued prayer and other spiritual practices will attract water and other energy features to a site.[22]

Just as negative geographic samskaras can create negative vortexes so can positive geographic samskaras create positive, or natural vortexes (NV) that can help uplift you. Feng shui advocates that we should build a home where chi (prana) accumulates. A natural vortex is a place where energy accumulates in a very positive and loving way.

As a positive geographic samskara gets stronger it can create a natural vortex of prana at a particular location. A natural vortex of prana creates a swirl of prana that better allows you to absorb it if you are in it or in close proximity to it. As the geographic samskara gets stronger it will eventually help create a natural vortex of cosmic prana. They generally form where a lot of praying and meditation have taken place and are great places to meditate if you want to raise your consciousness. They are a blessing from God and Mother Earth which give a great feeling of well being.

The prana drawn to a site enhances the space. When it combines with the spiritual consciousness of prayer it creates a powerful presence that can help raise your consciousness and facilitate connection to divine.

Energy lines and formations tend to remain in place for a very long time as a testament to the prayers that previously occurred there. It is good to pray at an old sacred site because you can rekindle and add to the foundation of love and energy built there. The increased energy can enhance your spiritual experience. We are drawn to the same site over and over again because of its energy and positive geographic samskaras that remain there. As new communities and civilizations develop they are often unconsciously drawn to build places of worship that were previously sacred sites.

The prana drawn to a sacred place, like Kundalini energy within the body, greatly enhances the spiritual nature of a sacred site.

When we meditate or pray in an area we not only give it strength but we begin to cleanse it of negative geographic samskaras. As the cleansing progresses more features of Mother Earth may be revealed. You may find that there is much more to a place than you previously thought as the blockages from the physical plane begin to diminish and divine aspects are drawn to it.

The Divine Light Of Meditation

When we love, have good thoughts, do good things, are focused, or perform spiritual exercises such as prayer and meditation we are in harmony with and connecting with source. We are also tapping into and absorbing all of Mother Earth's benefits and nutrients.

Meditation and spiritual exercises such as prayer take that connection a step further. When we meditate Mother Earth gives us the equivalent of a hug and we increase our draw of cosmic prana and other essences. When we meditate the flow of cosmic prana in the immediate area around us is pulled towards us creating a circle around us that can reach a few inches to several feet. This whirlwind of cosmic prana feeds our subtle body with increased consciousness. It is this dynamic interchange and intake of cosmic prana that underlies many of the benefits associated with meditation.

The space where we meditate also benefits as meditation will help cleanse and reduce negative geographic samskaras and plant a seed thought of attracting cosmic prana. That seed thought if reinforced with further meditation will create a natural vortex of cosmic prana.

Similarly, healing such as laying on of hands, therapeutic touch and reiki will draw earth prana and magnetic prana. Mother Earth truly responds to our intentions.

Connecting with Mother Earth

Living in harmony with Mother Earth is vital for our and Mother Earth's health and well being. Feng shui teaches that a myriad of benefits are derived when we live in balance with Mother Earth.[23] It places emphasis on where you should build your home and how it should be arranged to be in harmony and derive maximum benefits from a positive energy flow.

We can begin the process of healing Mother Earth by being cognizant of how our thoughts and actions affect her. We can also begin praying for her and choosing wisely where we pray. Sun Bear suggests that we should look for special places, what he calls power places, to pray: "I feel—and Spirit has also told me—that this is a time for human beings to find the spiritual beliefs that link them to the Earth. This is a time to seek power spots and find places to pray and reconnect with the natural forces and with the spirit helpers that have been upon this planet for thousands of years."[24]

While all of Mother Earth is sacred, certain locations that Sun Bear calls power places, can increase and enhance our spiritual experience. This may be attributed to the unique qualities of a place or because it has drawn a lot of Mother Earth's equivalent of kundalini energy.

When you intentionally decide to honor Mother Earth by praying at a sacred place or a place you wish to make sacred, you will be honoring her and showing respect. Praying at a sacred site will facilitate your connection to Mother Earth and enhance the location. In other words it will improve your sentience of Mother Earth and your ability to absorb her beneficial essences. A multitude of benefits will flow to you as you increase your sentience of Mother Earth; from raising your consciousness, to improving your health, to making you less vulnerable do diseases and contagions and it will make you feel very, very good. You will also begin the process of healing Mother Earth. Remember, we give strength to whatever we think about. Later in Chapter 12 "Sacred Earth—Creating Sacred Space" we will talk in greater detail about where we should pray and how to create sacred space.

Mother Earth Needs Love

Science is the not solution to ending global warming, pollution and the extinction of species. Science and the belief of the ego to overcome, along with violence and greed are the problems creating Mother Earth's diseases. Cutting back on pollution, using alternative forms of energy, manufacturing less, are noble and worthwhile efforts, but they are not going to solve the problem. We have damaged Mother Earth's subtle body to the point that she can no longer function properly. Global warming, increased disease and the extinction of species are the symptoms of the illness: our violent behavior and thinking.

We need to address the problem and heal Mother Earth's subtle body. We need to understand that the same mind-body dynamic at work within us is at work within Mother Earth, only it is our mind that is driving her body. Our collective mind is polluted. We need to cleanse it of violence, hate, envy, ego and materialism.

Mother Earth needs love, compassion and mercy from us, for each other and for all of nature. Mother Earth needs for us to do what the prophets have taught throughout the ages. The solution to healing Mother Earth is in our hearts, not in our minds.

Chapter 4

Inhabitants of Other Worlds

The unseen world around us, the various planes and sub-planes are full of inhabitants. They are constantly interacting with us and influencing us. Their nature runs from the divine to the demonic. Some of these inhabitants such as angels are very helpful, while others such as demons are very malicious. They play an important role in our development, a role that none of us truly understands.

The Illusion of Others

Some believe that there are no other beings existing in the other planes around us. To them any angel, spirit, or demon that we believe in, or possibly even see or feel, is purely imaginary. Many Buddhist's believe that all phenomena are illusions we create in our mind: "Within our mind there is a Buddha, and that Buddha within is the real Buddha. If Buddha is not to be sought within our mind, where shall we find the real Buddha? Doubt not that Buddha is within your mind, apart from which nothing can exist. Since all things or phenomena are the production of our mind."[1]

The Sutra of Hui Neng goes on to say that any Mara[2] (devil, demon) that we see is a figment of our imagination that ultimately must be transformed:

The Essence of Mind or Tathata (Suchness) is the real Buddha, While heretical views and the three poisonous elements are Mara. Enlightened by Right Views, we call forth the Buddha within us. When our nature is dominated by the three poisonous elements, We are said to be possessed by Mara; But when Right Views eliminate from our mind these poisonous elements, Mara will be transformed into a real Buddha.[3]

There is no doubt that our mind plays tricks with us and we can create illusions as well as other things. As was noted previously thought forms can take on life of their own. We can create thoughts that that can influence others, or enslave us. We can also create thought forms that can appear to be supernatural. For example, we can hear a creaking floor and say it is a spirit. We then begin to look for more creaking as a further sign of spirits, and so on and so on. Eventually the imaginary spirit takes on greater influence and the thought form itself may begin to create the creaking.

Alexandra David-Neel who brought Tibetan Buddhist and mysticism to the west early in the twentieth century in *With Mystics and Magicians in Tibet* tells what she calls a 'known and famous' story throughout Tibet about the power of the mind to create. In it a trader traveling with his caravan lost his hat on a stormy day and had it blown into some thorny bushes. A few weeks later another man saw the hat and later told the towns-people that he had seen something mysterious crouched in the bushes. Over time others began to take notice of the hat, particu-larly as the weather changed its color and contour. Then one day someone seeing its weathered form rustling from the wind commented that it was a demon. A few months later a group of villagers were sent off in a panic when they saw the wind blow the hat out of the thicket where it had been trapped. Alexandra David-Neel noted that, "The hat had been animated by the many thoughts concentrated upon it. That story, which Tibetans affirm to be authentic, is given as an instance of the power of concen-tration of mind, even when unconsciously affected, and not aiming at a prescribed result."[4]

More often then not it is our imagination that is creating an imaginary character. But there are inhabitants in other worlds.

Similarly, we create less ferocious monsters, things like superstitions with our group think. Sri Aurobindo's the Mother (Mirra Alfassa) says that superstitions begin when someone has an unfortunate accident after walking under a ladder or seeing a black cat. By talking about the experience to others we give more power to what happened; and if others pass along the story and further embellish what happened it can create, what The Mother calls a powerful 'mental formation.'[5]

It is important to remember that we attract what we think and those thoughts surround us. So if we are thinking of angels then it is angels that we will attract. If it is demons that we are thinking about then it is demons that we will attract. If we are thinking thoughts of the divine and how to help humanity and serve God then it is angels and divine beings that we will attract. If we have violent and vile thoughts then it will be demons and other dark beings that we will attract.

Artificial elementals, Tulpas, Golems

We certainly can and do create our personal demons. If we keep working a thought form long enough and with emotion we can create what Besant and Leadbeater call an "artificial elemental":

> The desire (or astral) body gives rise to a second class of entities, similar in their general constitution to the thought-forms already described ...Such a thought-form has for its body this elemental essence, and for its animating soul the desire or passion which threw it forth; according to the amount of mental energy combined with this desire or passion, will be the force of the thought-form. These, like those belonging to the mental plane, are called artificial elementals, and they are by far the most common, as few thoughts of ordinary men and women are untinged with desire, passion, or emotion.[6]

What Besant and Leadbeater describing with an artificial elemental is something beyond a thought form, or samskara. It is something that in a certain sense is an independent and living entity.

It is through thought forms that wizards, shamans and sorcerers often ply their craft. Some make a distinction between white magic, acts done for good, versus black magic acts done for power, violence or retribution. Practicing such arts is a very dangerous affair and many a wizard has met an early and tragic death.

In her studies with Tibetan mystics Alexandra David-Neel found that through concentrative powers, consciously and sometimes unconsciously, Tibetan mystics developed phenomena and apparitions. One form in particular, a 'tulpa', took on a life of its own and had great powers:

> Once the tulpa is endowed with enough vitality to be capable of playing the part of a real being, it tends to free itself from its maker's control. This, say Tibetan occultists, happens nearly mechanically, just as the child, when his body is completed and able to live apart, leaves its mother's womb. Sometimes the phantom becomes a rebellious son and one hears of uncanny struggles that have taken place between magicians and their creatures, the former being severely hurt or even killed by the latter.[7]

In his commentary on *Sefer Yetirah* Aryeh Kaplan notes how Kabbalahists were able to create what Jewish mystics refer to as a golem, or an artificial elemental. It could take over thirty hours of continually meditation and mantra to create a golem: "There is also evidence that creating a golem was primarily not a physical procedure, but rather, a highly advanced meditative technique. By chanting the appropriate letter arrays together with the letters of the Tetragrammaton, the initiate could form a very real mental image of a human being, limb, by limb. This could be used as an astral body, through which one could ascend to spiritual realms."[8]

In other cultures artificial elementals, tulpas and golems are known as a 'poltergeist.' Some new age types refer to them as 'egregore.' There are probably countless other names in other cultures that refer to unseen spirits created individually or collectively. The fact that so many cultures share a common belief in artificial elementals, tulpas and golems gives credence to their existence. The idea that a tulpa could be created unconsciously as Alexandra David-Neel notes is chilling.

Idols, False Gods

The creation of tulpas and golems and other such entities under-scores the power of our thoughts to create things beyond ourselves that affect the rest of humanity. If each of us has the ability to create tulpas and golems what could many of us, millions, or even billions of us create together? Our collective conscious and unconscious has a tremendous power to create all sorts of things.

That is the problem. Our thought forms can do more than become stronger and more domineering samskaras, they can become hideous creatures that can take on a life of their own. They can become false gods, or idols with a life and will of their own.

As with all thought forms false gods look to manifest their design, one that can twist and turn in diabolical ways. Similarly as with thought forms the power of thought applied to them adds to their strength. False gods need our thoughts and efforts for their sustenance.

Leadbeater notes how once, created entities such as tulpas can become demon-like and take control of groups of people and have them serve and care for it. They become false gods, or idols:

Such creatures occasionally, for various reasons, escape from the control of those who are trying to make use of them, and become wandering and aimless demons....They invariably seek for means of prolonging their lives, either by feeding like vampires upon the vitality of human beings, or by influencing them to make offerings to them; and among simple half-savage tribes they have frequently succeeded by judicious management in being recognized as village or family gods.[9]

The violence and bloodlust of the sacrifice make it stronger. The physical action of sacrifice reinforces the intent and the emotion (bloodlust) gives it emphasis and strength.

The world is full false gods and idols that dominate and manipulate our lives in a very pernicious ways, as will be detailed in Part II Hell, Our World, Our Making. We hesitate to bring God into our lives yet we allow ourselves to be continually manipulated by false gods.

Be cautious in your dealings

There are all sorts of beings close to us all the time. We are constantly interacting with them unknowingly. It is said that friendly spirits will respect your privacy, keep their distance, not sneak up on you and not trick you or take advantage of you. Conversely negative entities will sneak up on you, have no respect for you and are constantly looking to take advantage. So if you sense something close by look at their behavior as a possible sign of whether they are friendly or not.

I am not an advocate of trying to connect with inhabitants in other realms. A lot of people are constantly looking to meet nature spirits or divine creatures. They take walks in the woods hoping to connect with them. I strongly suggest that you do not make meeting divine creatures your purpose. First, your desire to seek can very easily be manipulated by something not so divine for its own purposes. This can be a very big problem because not only are you opening yourself up, but you are creating a thought form of connection that you will not able to control. Very dangerous. Secondly, you should respect the space of divine beings. When you hunt for beings in other realms you are no different than some half crazed tourist with a camera going all over the place to get a picture or some ATV rider going into the deep woods and disturbing animals. You may also be trampling in places that you really should not be in. Please respect their space.

Focus on God and developing your spiritual self first and foremost. If you do this all sorts of doors and encounters will be opened up to you. Not only will doors be opened but you will have developed discernment and the ability to cast off things you might not want to deal with. Meeting with divine beings such as angels is a good thing once you have developed your spiritual self, but let them come to you. Seek and give love and everything else will fall in place.

Nothing about us is hidden to others in the unseen world. Our thoughts are open and on display for all to read. The more sophisticated the entity the greater is its ability to read our samskaras, our history, and vulnerabilities. In a sense we exist naked in these other worlds because all is there to be seen. This information can be read and processed with amazing speed. It can be played back to you so that you believe that you are connecting with a deceased relative, or used to exploit your greatest vulnerabilities and fears.

Ghosts, Spirits

What many people call ghosts are the apparitions of dead people that have not passed along. It is not necessary to see an apparition to be in the presence of a ghost. As with thought forms, ghosts gain strength from the thoughts of others that are focused on them. An emotion tied to a thought adds strength to it and this explains why people often find that ghosts are looking to spook or scare them because it creates a much stronger thought form. The creaking doors and steps more often than not one hears if anything supernatural has a logical explanation and at most is a thought form. There are ghosts that are truly ghoulish and are looking to cause harm.

Many people look to connect with deceased relatives, but I would urge caution. Many mediums and channelers may in fact be reading your samskaras or even the samskaras of your deceased relative instead of communicating with your deceased relative. You also may be communicating with a fragment of your deceased relative's previous self, their lower physical and emotional self, and not their higher self that has split off.[10] If deceased relative wants to come to you it will do so, you only need to be open. Communication may come in subtle ways.

I would broadly classify all beings in other planes as spirits. Spirits are also those that have passed along and are waiting to reborn and if strong enough may communicate with you.

Pranic Plane

The pranic plane, or plane of energy, is Mother Earth's equivalent of our pranayama kosha and deals with energy and the maintenance of the subtle bodies. Inhabitants in this realm are concerned with energy and maintaining the lower planes.

There are two broad categories of inhabitants in the pranic plane. One are energy beings, generally negative entities, what many call demons; the other are nature spirits or what some call elementals, or faeries, nymphs, elf's, devas and a host of other names.

Leadbeater describes the pranic plane[11] as the plane of illusion because of the unreliability of what one sees there and the shape shifting ability of its inhabitants.[12] He notes the uncanny ability of its inhabitants to alter their appearance and how one can appear to see them from different angles simultaneously. You should always be cognizant of the shape shifting ability of entities when dealing with inhabitants of other planes. Caution is urged, most notably because negative entities can cloak themselves in light and try to deceive you.

Just as we have different personalities and proclivities so it is with inhabitants of the pranic plane, particularly those of a higher order. In other words, elementals of a certain type will vary in disposition. Some are friendly and helpful, others are not.

Inhabitants from other planes are able to travel long distances with great speed via spirit lines that cover the world. Spirit lines are lines of consciousness and are meant to carry God's highest aspirations for humanity. Like us, entities leave a bit of themselves wherever they go. So if demons are traveling along a spirit line they will leave a touch of their negative consciousness within it.

Negative Entities

The demons that Jesus exorcised and that are mentioned in the sacred texts of most faiths exist in the pranic plane. They generally are a malicious and mean spirited bunch that account for the bulk of the ill will done on the physical plane by inhabitants from other worlds. They torment humanity so that they can feed off of our vitality and energy.

There are also classes of beings who are pranksters, what some call the jokester archetype. These entities are playing the equivalent of practical jokes upon us, often doing so in a hurtful way.

Sri Aurobindo describes hostile beings as consisting of three types, with the Asura having the highest intellect, versus the Rakshasas and Pishachas, which are more like thugs. The Asura has no soul so it is not part of the evolutionary process. It has a very powerful ego and can be highly intelligent, but is driven by their desires. The Rakshasas are driven by violent passions and emotions.

They are the Powers of Darkness combating the Powers of Light. Sometimes they possess men in order to act through them, sometimes they take birth in a human body...The Asuras are really the dark side of the mental, or more strictly of the vital mind plane... Their main characteristic is egoistic strength and struggle, which refuse the higher law. The Asura has self-control, tapas and intelligence, but all that for the sake of his ego. There are no Asuras on the higher planes where the Truth prevails.[13]

Just as the physical plane (world) is inhabited with a multitude of creatures so is the pranic plane similarity inhabited with a myriad of energy feeders. Many of whom are neither violent nor particularly nasty.

You should be very careful about being part of a group that is creating a large amount of energy because it attracts beings focused on energy. For example, doing energy work, pranayama or other types of energy producing exercises, sweat lodges or other spiritual gatherings (worship, church, group meditation, etc.) can create an energy surge. Sadly energy feeders often dwell near a place of worship, or where spiritual practices are held. Mobs in particular can attract the negative, as can other emotional events such as sporting events. Demonic and predatory behavior such as war, killing, rape and hunting can attract negative entities. Understand that energy feeders are attracted to energy, particularly emotional and passionate events and gatherings.

To protect yourself you should always ask for protection and visually surround yourself in God's divine light and love.[14] Choosing your surroundings can help a lot in preventing unpleasant encounters. A place of worship that has positive geographic samskaras is a good start[15]; I would further suggest a place dedicated to God. Being devotional and focusing your practices on God, essentially doing them for God, will help thwart negative entities. Your spiritual strength, having a strong subtle body and soul, is your best ally. Make sure that whoever is leading a group can sense and knows how to deal with negative entities, particularly if you are doing energy work.

Energy, prana (chi), is a vital component in developing our spiritual self. At the same time it attracts energy feeders, presenting us with a challenge. Many problems the spiritual aspirant encounters are when the focus becomes solely on energy. Instead look to blend energy with consciousness (Godhead), Shiva with Shakti. Dedicate your practice to God.

Try not to be fearful should you ever encounter pranic beings, because fear gives them confidence. You should avoid confronting or challenging them, unless you are well schooled. They will get the better of you very easily. They can read your thoughts and possibly even your samskaras to know what motivates you and to know your vulnerabilities. They will also know what you are planning.

Many of the exorcisms Jesus performed were to remove a demon that had possessed someone, or had attached to them. In fact it appears much of what Jesus exorcised is what many call attachments. These are entities that attach to and drain you of your vital energy possibly causing debilitating diseases and even death.

Demons also play with our mind and speak to us about violence, negativity and the like. They have been known to manipulate people to disrupt the flow of things. If you hear voices speaking of violence or derisively you should immediately call the divine and focus on the positive. Demons can also plant negative seed thoughts on us that if we are not careful can blossom. Just because a demon or a thought form had us commit a sin does not exculpate us from the sin. The comic Flip Wilson in the 1960's and 70's had a nationally famous skit where he would parody the temptation of evil and would say," The devil made me to it." You did it or thought it, you own it. Something beyond

ourselves may have us commit sins but ultimately we are responsible for our own thoughts and actions.

The discussion of demonic behavior in modern society is not
to titillate, but to educate people about dangerous things. People
need to be told and to learn from this.

Nature Spirits

Nature spirits are the beings whose function is to interact with
and help maintain the physical world through a variety of
functions. They inhabit various sub-planes in the pranic plane.
Others broadly refer to them as elementals because they are the
most basic of beings that are early in their evolutionary process.

Generally speaking nature spirits have been generally known
to encompass beings such as faeries, devas, elves, leprechauns,
gnomes, pixies and the like. They have a mystical and magical
persona in the eyes of many people.

Nature spirits are a curious bunch, whose personalities, like
ours, can vary dramatically. Generally, they are cautious and
sometimes outright not very friendly, particularly the lower
nature spirits. I think this has a lot to do with their experiences
with mankind who has desecrated areas that they are meant to
maintain. We clear cut and burn forests, strip mine for coal and
stalk the woods looking to kill creatures in their domain. Why
should they be friendly given all this? Higher nature spirits such
as devas and the like are more hospitable and can understand
who and what we are up to as individuals.

By destroying the environment we have simultaneously
exterminated much of the nature spirits domain and may have
wiped out a good portion of them. This raises a host of questions;
can the environment be resuscitated without them and what role
do they need to play in revitalizing the world?

The idea of nature spirits is universal to all cultures. Cultures
separated by history and space all speak of short beings such as
elves and leprechauns. The Haudenosaunee of upstate NY

similarly talk of the Jo-ga-oh:

> Among the fable folk of the Iroquois, the Jo-ga-oh, or invisible little people are beings empowered to serve nature with same authority as the greater spirits. These little people are divided into three tribes, the Ga-hon-ga of the rocks and rivers, the Gan-da-yah of the fruits and grains and the Oh-dan-was of the underneath shadows.[16]

Gayaneshaoh (Harriet Maxwell Converse) goes on to note how the closeness of the Haudenosaunee to the land and the Jogaoh gave them knowledge to overcome physical hardships. She also notes on several occasions how the Jagaoh had helped the Iroquois. Such stories of help by the little people are universal to all cultures at a time when we lived in closer harmony with the earth and nature spirits.

My friend, Susan Wiener, who has a strong connection to the earth and nature spirits, says that we should bring an offering as a gesture whenever we enter a forest. She also recommends that we should ask for permission to enter and wait a few minutes before entering. Say a prayer before entering an area.

I find it helpful to treat as space as you would in making a friend. It takes time to develop a friendship and it is the same with knowing a space and its inhabitants. Treat your first visit as a getting to know someone and work from there. I often try and clean up a space by physically picking up litter and by spiritually cleaning it up. Show genuine concern and interest.

Angels

All faith traditions speak of divine beings, whether they are called angels, devas or minor Gods who are divine messengers of God. Angels have played and will continue to play a vital role in the development of our individual and collective consciousness. It was the angel Gabriele that communicated the whole of the

Quran to Mohammed. Gabriele told Mary that she would have a great son.[17] Angels of the lord appeared to Moses in the Burning Bush.[18] It was Angels that led the Israelites through the Promised Land.[19] Theosophist Geoffrey Hodson believes that there is a hierarchy of angels that like us is evolving and using the planet earth for their development.[20] In *Clairvoyant Investigations* Hodson describes angels as the glue and force that keeps the cosmos and various planes together:

> Angels or in Sanskrit "Devas", sometimes called Gods, are hierarchal orders of intelligences, quite distinctive from man....They are regarded as omnipresent, super physical agents of the creative will of the Logos, as directors of natural forces, laws and processes, solar, interplanetary and planetary...
>
> [W]hile themselves continuing to evolve, have a mission bestowed upon them, which they accept. This mission is single through the devic hierarchy and consists of responsibility for and continual assistance or "quickening" in the procedures of evolution.[21]

Similarly the Bhagavad-Gita speaks of divine intervention whenever there is decline: "Whenever and wherever there is decline of dharma (righteousness) and ascendance of adharma (unrighteousness), at that time I manifest Myself in visible form. For the protection of the righteous and the destruction of the wicked, and for the right of establishing dharma again, I incarnate Myself on earth from time to time."[22]

What the Gita is describing—'I manifest myself in visible form'—is what the Abrahamic faiths would call a prophet, or a divine messenger from God. I believe that many of the prophets were indeed angels. The actions taken by prophets to warn and foretell, to lead and to instruct are similar to the functions historically performed by angels.

To me things such as difficulty with speech, found with many prophets, is indicative of a higher being that is more at home with conceptual and telepathic communication rather than the word. Moses had trouble with speaking, it is rumored that Jeremiah had a secretary, and the Haudenosaunee prophet the Peacemaker had Hiawatha speak for him because of his speaking problems. Some of you might have noticed how speech can be a hindrance at times. For example, during meditation if you are concentrating on an object or concept your mental verbalization can impede and even set back your meditative experience.

While the Bhagavad-Gita notes that part of God transcends the physical word it does not mean that they are the only creatures of God, rather all is Braham. Angels are closer to the divine and farther along in their evolution.

Angels can manifest in the general populace or remain invisible and help from the sidelines. We can and should connect with them; however I would suggest letting them come to you rather than making contacting your goal. They are the vital link to God and to our individual and collective transformation. Divine beings are our best guides along the path. It is through spiritual exercises that we develop the spiritual self and the ability to communicate. These are vital skills that must be nurtured and developed. The Essene gospel of Peace says that we should connect with and strive to be like angels:

> Follow the example of all the angels of the Heavenly Father and of the Earthly Mother, who work day and night, without ceasing, upon the kingdoms of the heavens and of the earth. Therefore, receive also into yourselves the strongest of God's angels, the angel of deeds, and work all together upon the kingdom of God.... And your Earthly Mother and Heavenly Father will send you their angels to teach, to love, and to serve you. And their angels will write the commandments of God in your head, in your heart, and in your hands, that you may

know, feel, and do God's commandments.[23]

Wise words.

Let the disturbing description of negative entities like demons serve as a cautionary tale. Yet if we approach the divine in earnest and with the most sincere and innocent intentions and with the utmost faith we will be protected. Remember the words of Jesus who said upon performing each miracle that "your faith has saved you." Similarly if we look to the miracles of Fatima, Chimayo and Lourdes it was the innocent, the sincere and the faithful that saw and were healed. Be like them and divine beings will take you under their wing.

Part II Hell: Our World, Our Making

Chapter 5

Seed Thoughts and Their Fruit

While we live in a mystical world we have trapped ourselves in the material world. We are not free. This material world, or physical plane, is dominated by our thoughts that have found an independent existence in the mental plane. Our thoughts have created false gods who look to control us and get us to worship them and thereby strengthen them. If we are to liberate ourselves we need to learn how our thoughts have morphed and enslaved us. The following details how we how we have trapped ourselves.

Thoughts Morph

Each thought that we have gives strength to, arguably feeds whatever we are thinking about. At the same time we are planting a seed that will try to bear fruit by driving us to repeat the same thought pattern. Once a thought starts to gain strength from repetition it also begins to evolve, or metamorphosize into something much more powerful. The Bhagavad-Gita describes how once planted a seed thought blossoms:

> While contemplating the objects of the senses, a person develops attachment for them, and from such attachment lust develops, and from lust anger arises. From anger, delusion arises, and from delusion bewilderment of memory. When memory is bewildered, intelligence is lost, and when intelligence is lost, one falls down again into the material pool.[1]

The idea that attachment can lead to lust and then anger demonstrates the power of thoughts to evolve. A seed thought once planted can mutate into many things; anger instead of leading to delusion can lead to hate, hate to violence, violence to murder. That is the challenge.

To understand how our thoughts can mushroom into something more complex we need only think of rumor. All of us at some point have fallen victim to rumor. There is the story of the wife that met with her husband's best friend in secret to discuss a surprise birthday party for her husband. Mutual friends seeing the two remarked about the secrecy and the unusual nature of the meeting. They told others who embellished the story with their own take. Eventually the husband is told that his wife is having an affair with his best friend. Unfounded suspicions as we heard with the Tibetan hat story by Alexandria David-Neel can build momentum of their own and give existence to something beyond themselves.

Swept Away By Our Thoughts
The Bhagavad-Gita notes that the power of the material world is so strong that it can carry away even the most ardent and devoted disciple, "The senses are so strong and impetuous, O Arjuna, that they forcibly carry away the mind even of a man of discrimination who is endeavoring to control them."[2]

In his commentary of the verse Swami Prabhupada notes how easily swayed are those not in Krishna(God) consciousnesses: "There are many learned sages, philosophers and transcendentalists who try to conquer the senses, but in spite of their endeavors, even the greatest of them sometimes fall victim to material sense enjoyment due to the agitated mind....Therefore, it is very difficult to control the mind and the senses without being fully Krishna conscious"[3]

What Swami Prabhupada is saying is that where our consciousness resides, or where we plant our seed thoughts will determine where our thoughts go. If we embrace God—through thought and action— then our chances of transcending the material world greatly increase. If we plant our seed thoughts in sense objects and desires then we sink deeper into the quagmire of delusion, ignorance and suffering.

The problem is that we, humanity, are currently tethered to the physical plane of sense objects and desires. So just about every seed thought that we have is doomed because it is focused on material objects or desires. Thinking about God continually can be challenging.

Seed thoughts are like the foundation of a house upon which we build. That first thought and its intention will determine the direction that our thinking goes. That is because that first thought will look to bear fruit and plant more like minded thoughts. Over time the foundation will get stronger and stronger from continual thinking. It will also begin to mutate and evolve in unimaginable ways. Eventually what is on the foundation, while born of it, may now look dramatically different because of the continual permutation of thoughts over time.

Where we plant our seed thoughts as a society will similarly determine the fruit that it bears. If we plant seeds of love, peace and God then there is a good chance that they will bear loving and divine fruit. Similarly if we plant seeds of personal gain of winner take all, then it will bear the fruit of ruthlessness, corruption, violence and worse.

Unfortunately, we have built many of our values upon a winner take all society believing that the winnowing process of competition benefits us all. The premise being that the process of competing makes us better, improves our well being, rewards the diligent and encourages the losers to work harder. To substantiate the benefits of competition pundits point to triumphs in sports, medicine, science and business.

Darwin Substantiates the Power of Violence

Science heralds Darwin's theory of evolution, survival of the fittest and the process of natural selection, as proof positive that competition is what makes us the success that we are. In reality Darwin's theory of survival proves that those best able to kill, most willing to exploit and disregard anything or anyone else besides themselves advance in a physical plane based upon self interest. According to Darwin we must put niceties aside when looking at species, including humanity, and see that we live in a dog eat dog world where we all feed upon each other:

> We behold the face of nature bright with gladness, we often see superabundance of food; we do not see or we forget, that the birds which are idly singing round us mostly live on insects or seeds, and are thus constantly destroying life; or we forget how largely these songsters, or their eggs, or their nestlings, are destroyed by birds and beasts of prey; we do not always bear in mind, that, though food may be now superabundant, it is not so all seasons of each recurring fruit.[4]

Darwin notes that those species with the competitive edge are favorable and those that do not are injurious. He calls this weeding out process natural selection: "The preservation of favorable individual differences and variations and the destruction of those which are injurious, I have called Natural Selection or the Survival of the Fittest."[5]

According to Darwin the same dog eat dog world of plant and animal species applies to man as well,

> We have now seen that man is variable in body and mind; and that the variations are induced, either directly or indirectly, by the same general causes, and obey the same general laws, as with lower animals. Man has spread widely over the face of the earth, and must have been exposed, during his incessant migration, to the most diversified conditions.[6]

In describing those characteristics that led to man's triumph Darwin emphasizes violence and killing:

> He manifestly owes this immense superiority to his intellectual faculties, to his social habits, which lead him to aid and defend his fellows, and to his corporeal structure. The supreme importance of these characters has been proved by the final arbitrament of the battle of life... He has invented and is able to use various weapons, tools, traps, etc. with which he defends himself, or kills or catches prey, or otherwise obtains food. He has made rafts or canoes for fishing or crossing over to neighboring fertile islands. He has discovered the art of making fire..."[7]

Science's Creation Story of Violence

The idea that competition among species is the heart of who we are is pure hogwash. Competition among species was generated by mankind's thoughts of self-interest going back eons and eons ago. Darwin's theory of evolution and survival of the fittest shows how a seed thought of violence and self interest once planted blossoms to triumph by promoting and advancing those most able to adhere to this principle. So if we plant seeds thoughts of self-interest and violence then the successful will be the most devoted to it through thought and action.

Darwin's theory of natural selection stands in direct opposition to God's call to love one another and the message and actions of the prophets to stand up for the pariah and disadvantaged. Instead of embracing God and planting seeds of love we have planted seeds of self-interest and competition. When self-interest and competition become the foundation of our thoughts it will ultimately bear fruit of violence, exploitation, hate, rape, murder, war, genocide, ethnic cleansing and other forms of violence.

Darwin's theory of natural selection showcases the dogma of

science and materialism. It is science's and the materialist's creation story. It is science's equivalent of the bible's Genesis; only instead of having God creating a peaceful world and a Garden of Eden Darwin's creationism tells of the triumph of wickedness and violence. It substantiates and gives credence to violence and the abuse of others.

Walter Wink believes that it is the primordial myth of violence that we have created in our society that shapes our current ethos of violence.[8] Wink is fond of telling how the Babylonian creation story creates a culture of violence by having a captive God murdered and then using its blood to create humanity. He asks; "If human beings are created from the blood of a slaughtered god, how can one expect from them anything but violence?"[9]

Similarly Darwin's creation story says that violence and murder is what we are about and that it is good because killing is what makes us successful. Holding such a model up inculcates us, especially the young in schools, to accept violence and believing that winning justifies doing whatever it takes.

Competition's Wicked Fruit

The malicious fruit borne by the seed thoughts of competition are all too apparent. Competition leads to winning at all cost, to the disregard of the law and ultimately to violence. Competition begins with the seed thought of beating out your competitors. Attention is focused on how to win, how to beat your opponent and what lengths to go to achieve victory. Even the language of competition is derisive and violent—enemy, opponent, winner, and loser.

The microcosm of sports is symptomatic of the larger problems created by competition in society. What seems like innocuous competition between two teams, or individuals, ratchets up and up. Participants begin breaking the rules and sportsmanship falls to the wayside. Cheating and rule breaking has become so common that Dave Anderson of the NY Times noted, "In a sports year unlike any other, 2007 too often meant scandal and sadness rather than success."[10]

In December 2007 the Mitchell report confirmed what had been rumored for over a decade, that baseball had been rife with players illegally using drugs to enhance their performance. Not only were baseball players intentionally breaking the rules but there appears to have been several cover-ups and blatant disregard for the rules of baseball and the law by many higher ups as the NY Times reported; "The report was critical of the commissioner's office and the players' union for knowingly tolerating performance-enhancing drugs. It cited many instances where club officials knew about particular steroid use among players and did not report it. 'There was a collective failure to recognize the problem as it emerged and to deal with it early on,' Mr. Mitchell said."[11]

Mitchell was concerned that hundreds of thousands of high school athletes were also abusing and illegally using performance enhancing drugs. The illegal use of steroids was not confined to baseball. Olympic Gold Medalist Marion Jones admitted to steroid use and had all her records expunged. Similarly Ken Landis winner of the Tour De France bicycle race in 2006 was stripped of his medal over steroid use.

While cheating from NASCAR racing to basketball has been a feature of professional sports for years, it was shocking for many to learn that football's champion New England Patriots coach Bill Belichick was caught illegally spying on opponents in September 2007. It was reported that Belichick may have broken rules and spied on teams in previous years.[12] One must wonder if any of the Patriots super bowl wins were achieved by rule breaking. But no one really cared about the law breaking because the Patriots had a record undefeated season in 2007 and Belichick was named coach of the year for 2007 by the Associated press. What message did the world of sports send when it name Belichick coach of the year and made Tom Brady the team's quarterback player of the year ?

The disregard of the law and rules is not confined to sports. Similarly business which is based upon aggressive competition has a long history of law breaking—Enron, Equity Funding, Sawtek, Worldcom, Bernie Madoff among others.

Legal Band-aids Don't Work

When we plant seed thoughts of competition and self-interest in society they will bear fruit of the same ilk that will get increasingly stronger. To prevent ourselves from being over run by violence society passes laws. But what are laws? They are nothing but band aids meant to stop something. As we noted earlier whenever we are against something we give strength to whatever we are against. The heart of Tantric thinking is that we cannot stop our base behavior and need to put forth and embrace a positive alternative.

While the American constitution contains some lofty ideals and principles its intention rests in the material world and does not embrace God at its core. To maintain those lofty ideals espoused in the constitution we have been forced to create laws upon laws in hope of taming the worst in us. We have also had to create institutions and organizations to maintain those laws. The farther we get away from our ideals the more laws that we have had to pass in hopes of resurrecting them.

When we create a law, like with all thought forms, we plant several seeds simultaneously. One of which is adding to the process of, or pattern of, law creation. In other words there is a thought form associated with law creation that deals with the process of creating laws and looks to create more laws and in essence more bureaucracy. Justice becomes 'the system' with a life and set of rules all its own. In the process justice gets farther and farther removed from its original purpose.

Legal scholar Thane Rosenbaum notes that there is a

difference between justice today and doing what is right because the legal system has become institutionalized. Justice lives in its own little world of rules and regulations. He feels that, "Justice in many ways, has far more in common with the soulless, airless atmosphere that Franz Kafka concocted for this his character, Joseph K. in The Trial, than anything that approximates just treatment or a just result at the end of a long trial...Kafka's portrayal of justice is horrific, but perhaps all too accurate."[13]

Embrace the Divine

Jesus, as noted in the Bhagavad-Gita, believed that we should first and foremost embrace God:

'Teacher, which commandment in the law is the greatest?' He said to him, 'You shall love the Lord your God with all your heart, and with all your soul, and with all your mind. This is the greatest and first commandment. And a second is like it: You shall love your neighbor as yourself. On these two commandments hang all the law and the prophets.'[14]

Jesus is telling us to incorporate God in all that we do and think. That in order to transcend the material world we have to establish ourselves in God. He also points out that the second commandment to love your neighbor is related to our love of God and that all laws hinge upon and evolve from the two seed thoughts. That is the foundation.

The idea we should embrace God in government and in all that we do may seem too radical and frightening to many conjuring up visions of a feudal and archaic theocracy. Others may feel that the separation of church and state is vital. However, there is a sharp distinction between religion which is an institution and spirituality which is consciousness as we shall discuss throughout.

We need to remember that it was God's prophet, the

Haudenosaunee's Peacemaker that gave us the Great Law of Peace, upon which the American model of democracy is based[15] and became a model for others. In the Great Law of Peace the Peacemaker says that to have peace we must always keep the Creator in our mind: "Hearken, that peace may continue unto future days! Always listen to the words of the Great Creator, for he has spoken. United people, let not evil find lodging in your minds. For the Great Creator has spoken and the cause of Peace shall not become old. The cause of peace shall not die if you remember the Great Creator."[16]

Unfortunately the founders failed to incorporate the embrace and thought of God in the constitution.

Where we plant the first seed thought will determine its foundation and course that a thought form will take. Unless that seed is planted within God (Source, Spirit, Creator) or love it is going to take us down the wrong road and there are no laws or other measures that we can take to prevent that from happening. Plant the first thought seed in God.

The Rule of Larger Thought Forms

There are all sorts of thought forms that have been around for centuries and even eons that continue to exert an influence upon us; some of them are science, greed, violence, others are good like our thoughts of God and love. All vie for attention and look to influence. Since thoughts can mushroom into something bigger and different it is difficult to say what began when. For example, when did science begin? Did it begin with the renaissance and with the likes of Sir Isaac Newton, or did it begin earlier with Aristotle, or perhaps even earlier, even as early when mankind mistakenly accepted the reality of the maya of the physical plane? Then there are offshoots, technology that was the birthmother for electronic and eventually computer technology. There is biology, chemistry, mathematics, physics all offshoots of science. Biology has evolved into specialties such as genetics, ecology, physiology, medicine, etc...

What is not confusing is that all thought forms dealing with the physical plane (not God or divine qualities) drag us farther away from God and make us more dependent upon them (the thought form). In doing so they exert more influence on us, creating a dependence of sorts, and have us bear fruit of more like minded thoughts.

The Delusion of Science

We live in the age of science, technology and the rational mind. But science is neither the ultimate reality nor does it provide a path for liberation and happiness. Science is a very large and powerful thought form, a false god that has grown and morphed over a long period of time. Like all thought forms it looks to try and bear fruit and plant more seed thoughts. In the process we have become addicted to and dependent upon science and technology.

James Burke and Robert Ornstein in *The Axemaker's Gift—A Double Edged History of Human Culture* tell how supposed advancements in technology throughout history created a double edged sword of sorts, bringing good with bad and often with the bad outweighing the good. According to Burke and Ornstein the tools or technological innovations created by who they call "axemakers", changed society and forced changes upon their respective cultures, even altering a culture's way of perceiving and thinking.[17]

Each new technology changed the course of history and altered the balance of power and often created new structures and movements:

The effect of Gutenberg's letters would be to change the map of Europe, considerably reduce the power of the Catholic church, and alter the very nature of the knowledge on which political and religions control was based.

The printing press would also help to stimulate nascent forms of capitalism...

The political result of these new print-languages, imposed by kings through their control of the presses would lead directly to the emergence of a new kind of patriot axemaker...a nation.

Monarchs and their governments now began to enforce the local tongue with laws, taxes, armies and the state bureaucracies that went with them all....

The potential of the press for extensive bureaucratic control did not escape the attention of governments.[18]

Burke and Ornstein note that the rate of innovation and power of the axemaker's increased over time. Technological innovation increasingly brought deleterious consequences such alienation, social and economic inequality and increased the ability of the powerful to manipulate the masses:

Throughout history, mysterious axe maker knowledge always strengthened social conformist as at the same time it increasingly distanced the change-makers and their institutional masters from the general public whose lives they controlled. The sheer scale and number of new control systems generated by late-eighteenth century technologists and entrepreneurs widened this gulf and imposed rigid conformity as never before. Such was the rate of industrial innovation that it would force sudden and fundamental change on a society politically and administratively unready to deal with them. The changes would in turn bring into being new ways to manipulate the proletariat, because thanks to the factories there was a proletariat to manipulate. The new gifts would be an ideological tool for control...

The new industrial towns cut off the new village immigrants from nature and from any regard for it.[19]

What Burke and Ornstein are describing when they talk about tools and technology is a powerful thought form (talisman) that has built over many eons by our collective thoughts. The focus on control and overcoming Nature and Mother Earth for so long has increasingly brought forth new innovations to accomplish these ends. As with all thought forms the material drive of technology looks to increase its grip over us, by alienating us and making us more dependent upon it—for example creating factories that created a proletariat that could be manipulated. All this thinking and focus on a thought form is akin to worshipping it. Technology's hold on us is so strong that it would be nearly impossible to live today without it. How this thought form began and what amalgamations and unions it has made over time is next to impossible to ascertain.

Not only has science corralled mankind but it has ravaged Mother Earth and in doing severed a vital relationship to our health and spiritual evolution.

The New Reality

The process of scientific thinking of separating the mind from the body, spirit from matter, has increasingly alienated us from nature (Mother Earth) and deluded our perception about our true nature. David Suzuki believes that a few seed thoughts planted by a handful of 'thinkers' changed the course of history by getting us to think in certain ways that we today take for granted;

> Many thinkers trace the origins of our particular and violent fall from grace, our exile from the garden, back to Plato and Aristotle, who began a powerful process of separating the world-as-abstract-principle from the world-as-experience— dividing mind, that is, from body, and human beings from the world they inhabit. In the process they laid the groundwork for experimental science.
>
> Through Galileo, who identified the language of nature as

mathematics (an abstract language invented by humans), and Descartes, who learned to speak that language powerfully, the modern world emerges. Descartes famous definition of existence ("I think, therefore I am") completes a new myth about our relationship to the world: Human beings are things that think (the only things, and that is all they are), and the rest of the world is made up of things that can be measured (or "thought about"). Subject or object, mind or body, matter or spirit: this is what the dual world we have inhabited ever since—where the brain's ability to distinguish and classify has ruled the roost. From this duality come the ideas we live by, what William Blake called 'mind forg'd manacles', the mental abstractions that seem too obvious to question, that construct and confine our vision of reality.[20]

Science has gotten us to accept its dogmas as reality. This is a profound notion because it shows how a few misplaced thoughts over time can shape the perception of reality for generations to come. By accepting science's view of reality we have built more and more thoughts and artifacts upon its foundation. In doing so we have made science stronger and dragged ourselves farther and farther away from our true reality, that we are one.

As the false god of science has grown and evolved it has increasingly shut out competing alternatives that it vies with to occupy our minds. Huston Smith in *Why Religion Matters* says that one of the ways it has done this has been by secularizing and removing God from our universities, most of which began as religious institutions:

The most important cause of the increased secularization has been the progressive 'technologizing' of the Western world in the name of progress, and universities have been key agents in that project. Scientists needed to discover new laws of nature, and engineers to put those laws to use. Everybody got into this act, not just universities and scientists, for from healthy bodies to microwave ovens to television sets, material goods are the most obvious trophies that life sets before us.[21]

He goes on to detail the sweep of science in secularizing and dehumanizing universities while advancing the scientific perspective and related disciplines.

Global Monkey Brain

Our increasing reliance on science and technology has made us more vulnerable, pliable and distant. We are living in a world of stimulus overload. We are bombarded by electronics, cell phones, radios, and televisions, Ipods; by advertisements and by our jobs. To deal with all this input we do several things simultaneously and multi-task. This increased stimulus creates a global monkey brain that makes us individually and collectively unable to concentrate or feel at peace.

Monkey brain describes the mind of a monkey that jumps from thought to thought in quick succession. A monkey brain limits our ability to concentrate and focus. When we begin to learn to meditate we see that our mind jumps around like a monkey. To advance in meditation it is necessary to still the mind. Similarly a global monkey brain tears at our collective mind, we cannot focus, we cannot be still, we cannot see God. We are scattered.

Dr. Jon Kabat-Zinn Head of the Center for Mindfulness in Medicine, Health Care and Society at the University of Massachusetts Medical School feels that this process of technologizing our psyche and its ensuing disconnect has accelerated;

[M]uch has changed for us in the last hundred years, as we have drifted away from intimacy with the natural world and a lifetime connectedness to the community into which we were born. And that change has become even more striking in the past fifteen or so years, with the advent and virtually (pun intended) universal adoption of the digital revolution. All of our 'time-saving' devices orient us in the direction of greater speed, greater abstraction, and greater dis-embodiment and distance. It is now harder to pay attention to any one thing and there is more to pay attention to. We are easily diverted and more easily distracted. We are continuously bombarded with information...These assaults on our nervous system continually stimulate and foster desire and agitation rather than contentment and calmness.[22]

This stimulus overload has us frantically responding to it rather than taking a step back to see what is really going on. Kabat-Zinn says that this forces us to get us more into our heads and has us increasingly interfacing with machines that are neither alive nor natural.

Our brains and thinking process have been basically 'rewired' to suit the needs of technology and science. This reformatting has made us more dependent and more pliable to manipulation. It has also reduced our self-defense mechanism for escape. Because we cannot pay attention we cannot concentrate. Concentration is vital to meditation and meditation is a path to liberation as Buddha and Patanjali teach us. Concentration and meditation teach us about our thoughts and discernment. When we practice mindfulness or are meditating we are tapping into Mother Earth's vital essences that sustain us and make us feel good. We are also absorbing cosmic prana and raising our consciousness. The increasing necessity of interface with machines not only increases our dependence on them and our alienation from others and the world, but it robs us of our maternal relationship with Mother Earth. When we are distracted and multitasking we are not connected to Mother Earth and are subsequently deprived of her essences and making ourselves vulnerable to disease.

Dr. Kabat-Zinn continues by asking whether diseases such as A.D.D. (Attention Deficit Disorder) and A.D.H.D.(Attention Deficit Hyperactivity Disorder) are not a normal response for children entrained by adults into distraction and hyperactivity.[23]

Chapter 6

Collective Thoughts and False Gods

The idea that similar thoughts are attracted to each other and can form a contagion of like minded thoughts is a very frightening realization. As was noted in the last chapter one thought can create a strong samskara that can influence us, arguably control us, and a multitude of thoughts from a multitude of people can over time exert great control.

A contagion of like minded thoughts can also coagulate to become what is arguably a separate entity unto itself; what was described earlier as a tulpa, golem or artificial elemental. Such an entity's nature, motivation and purpose can vary from divine to demonic. It would, as with all of our thoughts, look to exert its influence and manifest itself and in the process could mushroom into something much, much bigger. In other words it would look to influence and control us. Such an entity would be a form of idolatry, or a false god.

Group Think

Whenever a group of people gather it has the ability to create a being unto itself. Irving Janis who has done a lot of work on 'groupthink,' the process whereby people abandon their critical thinking and analysis to achieve consensus for the group notes:

Groups like individuals, have shortcomings. Groups can bring out the worst, as well as the best in man. Nietzsche went so far as to say madness is the exception in individuals but the rule in groups. A considerable amount of social science literature shows that in circumstances of extreme crisis, group contagion occasionally gives rise to collective panic, violent acts of scapegoating, and other forms of what could be called group madness. Much more frequent however, are instances of mindless conformity and collective misjudgment of serious risks, which are collectively laughed off in a clubby atmosphere of relaxed conviviality.[1]

Janis compares groupthink to George Orwell's *1984*:

I use the term 'groupthink' as a quick and easy way to refer to a mode of thinking that people engage in when they are deeply involved in a cohesive in-group, when members' strivings for unanimity override their motivation to realistically appraise alternative course of action. 'Groupthink' is a term of the same order as the words in the newspeak vocabulary George Orwell presents in his dismaying 1984—A vocabulary with terms such as 'doublethink' and 'crime think'. But putting groupthink with those Orwellian words, I realize that groupthink takes on an invidious connotation. The invidiousness is intentional. Groupthink refers to a deterioration of mental efficiency, reality testing, and moral judgment that results from in-group pressures.[2]

The idea that forming a group makes the union invidious or obnoxious seems to conflict with the unity and oneness taught by God. This is because of the focus on things other than divine; seed thoughts not planted in God as was noted in the previous chapter leads to problems.

One of the negative consequences of groupthink is to treat the nonconformist as a pariah:

Attempts to influence the nonconformist member to revise to tone down the dissident ideas continue as long as most members of the group feel hopeful about talking him into changing his mind. But if they fail after repeated attempts, the amount of communication they direct toward the deviant decreases markedly. The members begin to exclude him, often quite subtly at first and later more obviously, in order to restore the unity of the group.[3]

In other words, the survival and sustenance of the group is put above friendship and compassionate treatment of others. If we assume that larger amalgamations of people exhibit group think then it becomes apparent why we so easily castigate the outsider and the different one in our society.

Similarly, herd behavior, when people move together in lockstep doing the same thing like a herd of cattle, has long been seen in the financial markets. Herd behavior occurs in financial markets when the buying and selling of traders is being driven by the buying and selling of others (the group). Ex Federal Reserve chief Alan Greenspan in 1996 called the US stock market's overvaluation 'irrational exuberance'[4] to describe how investors were abandoning reason and conventional valuation norms and kept buying stocks; they were buying because everyone else was buying. On its *Contagion of Financial Crisis* webpage[5] the World Bank found herd behavior increased during financial panics.

The Mob

Gustave Le Bon observed that organized groups of people have long played a major role in influencing events. Crowds of people numbering in the thousands or more are constantly gathering and forming in public places. At times these crowds, or groups, can take on a life of their own as individuals loose their own personal identity and associate with the larger collective. Le Bon found that there are moments when a triggering mechanism such as violent emotions associated with a great national event creates what he calls a 'psychological crowd.'

What turns a group of people into a 'psychological crowd' is the formation of a collective mind that becomes a separate entity unto itself.

> Whoever be the individuals that compose it, however, like or unlike be their mode of life, their occupations, their character, or their intelligence, the fact that they have been transformed into a crowd puts them in possession of a sort of collective mind which make them feel, think and act in a manner quite different from that in which each individual of them would feel, think and act were he in a state of isolation. There are certain ideas and feelings which do not come into being, or do not transform themselves in acts except in the case of individuals forming a crowd. The psychological crowd is a provisional being formed of heterogeneous elements, which for a moment are combined, exactly as the cells which constitute a living being which displays characteristics very different from those possessed by each of the cells singly.[6]

The crowd is not necessarily the mob in the town square. Crowds can link people living miles apart or in different countries. It is the collective mind, or focus on one thought or ideal, that connects them. For example, in the immediate aftermath of the 9-11 terrorist acts a horrific event that shook the world, a national, arguably global crowd formed focused on justice and sympathy. People cast their differences aside, all people became Americans and all Americans became New Yorkers. It also instilled great emotions, patriotism and self sacrifice.

According to Le Bon the psychological crowd is governed by the unconscious and is only capable of performing simple unsophisticated acts. The crowd is impulsive, easily irritated and its members are incapable of independent thought. Each individual of the group is caught in a hypnotic trance-like state that makes them susceptible to all sorts of influences and makes them malleable.[7]

The trance-like state that Le Bon describes seems to be like some sort of meditative state. Mystical experiences are associated with the divine, but as Le Bon tells us a group can be, "generous or cruel, heroic or cowardly, but they will always be so imperious that the interest of the individual, even the interest of self-preservation, will not dominate them."[8] More often than not in our current state of affairs groups are associated with violence and mobs, but later we will show how the collective mind can greatly enhance one's spiritual experience.

Groups Lose the Individual Sense of Self

Le Bon found that crowds have been known to collectively see illusions and to be easily swayed; "[S]uggestions are contagious....[A] crowd, as a rule, is in a state of expectant attention, which renders suggestion easy."[9] It is a group's malleability that makes them particularly vulnerable. For example, in the immediate aftermath of the 9-11 terrorist acts the Bush administration seized upon that moment to push through a radical agenda of laws that defied the American constitution[10] and values to invade Afghanistan and Iraq, all not in keeping with the American spirit of justice. A previously bickering Congress passed the Patriot Act and other measures that allowed increased surveillance of American citizens. The loss of self within a group also leads to the loss of responsibility and the consequences thereof: "The violence of the feelings of crowds is also increased, especially in heterogeneous crowds, by the absence of all sense of responsibility. In crowds the foolish, ignorant, envious persons are freed from the sense of their insignificance and powerlessness, and are possessed instead by the notion of brutal and temporary strength."[11] The fear is that the crowd can become an unruly mob bent on violence as we have seen with lynchings, riots and the like.

Hooliganism, violent and destructive behavior at sporting events has become all too common. What appears by many to be a little fun between two opponents is all too often a mini war. As Allen Guttmann writing on 'Sporting Crowds' notes:

> In sporting contests, more than in most social situations, the collective self is clearly defined against the collective other. This fact helps to explain why sports generates fan ship that is more intense, more obtrusive, and more enduring social activities. It is the intrinsically agnostic character of spectator sport that has always, from antiquity to modern times, made them especially suitable for their representative function. For sports fans, the appeal of the contest is that those who represent us block, tackle, kick, punch, pummel, or pin them, whoever they are. And because it is 'just a game' we can reassure ourselves, once we have calmed down, that our emotional binges are harmless.
>
> The problem is that the binges—the intense identification—are not harmless. They increase the fans propensity to behave aggressively. There is always the danger that partisanship will become hostile and that hostility will take physically violent forms. The historical record demonstrates that the danger is not trivial.[12]

The violence arising from a sporting event that Guttmann talks about is not confined to the stadium area or to the spectators. Whenever ever we think about something intensely and with great emotion we give it added strength. When we do it as group we are creating a powerful thought form that can become a false god. One that will be unleashed to manifest as Guttmann says by trying to have us 'block, tackle, kick, punch, pummel, or pin' others physically, or do so violently in our minds.

A Greater Purpose Drives Crowds

Le Bon felt that religious sentiment is at the heart of the crowd. He is not necessarily thinking of religion in the traditional sense.

What makes it religious according to Le Bon is the unswerving sacrifice of its members to put aside their own essence and being and wholeheartedly endorse the crowd's credo, regardless of whether it is divine or demonic:

> The sentiment has very simple characteristics, such as worship of a being supposed superior, fear of the power with which the behind is credited, blind submission to its commands, inability to discuss its dogmas, the desire to spread them and a tendency to consider as enemies all by whom they are not accepted. Whether such a sentiment apply to an invisible God, to a wooden or stone idol, to a hero or to a political conception, provided that it presents the preceding characteristics, its essence always remains religious. The supernatural and the miraculous are found to be present to the same extent. Crowds unconsciously accord a mysterious power to the political formula or the victorious leader that for the moment arouses their enthusiasms.
>
> A person is not religious solely when he worships a divinity, but when he puts all the resources of his mind, the complete submission of his will, and the whole-souled ardor of fanaticism at the services of a cause or an individual who becomes the goal and guide of his thoughts and actions.
>
> Intolerance and fanaticism are the necessary accompaniments of the religious sentiment.[13]

Le Bon cites the Jacobins of the Reign of Terror and the Catholics of the Inquisition as examples of how conviction of all sorts can link people. While the altars are long gone, new means and methods, new statues, portraits, dogmas, identities and cults have developed to sway participants in a group. Because "[t]he crowd demands a god before everything else."[14] Our modern world is full of similar false gods that people worship.

The necessity for a crowd, or any group of people called to

sacrifice, to have a God is why leaders are always invoking the 'God is on our side' during war. President Bush conjured up the battle between good and evil and invoked God to justify the invasion of Afghanistan and Iraq.[15] Similarly Osama Bin Laden played on ancient hatreds of crusaders and called for Jihad in an epic battle to save Islam and the holy lands. Like others before him Bin Laden called upon radical imams to provide him with a friendly interpretation of the Quran to help justify his defiance of Mohammed's teachings. Augustine concocted the 'Just War Theory' to enlist in the fight against the barbarian invasions of Rome during the fourth century. Violence and killing in the name of God is never justified. The problem is that these usurpers have left an indelible negative mark on religion and God.

For the atheists and agnostics it has been 'the cause' that has been the battle cry which justified self sacrifice and violence. Bolsheviks, violent revolutionaries and others have long held up 'the cause' as their defining purpose to motivate their followers.

Worship, Sacrifice and Violence

The crowds, or mobs, that Le Bon describes are more than a thought form that looks to dominate our thinking and control us, it demands that we worship and sacrifice to it. It is a false god; A separate entity. The idea that we would loose our own identity and possibility give our life to something is what makes it a false god. A false god is much more malevolent than a thought form. The creation and attributes of a crowd that Le Bon describes is very similar to the creation and attributes in making of an artificial elemental, or tulpa. The fact that a tulpa was intentionally created and a crowd forms spontaneously, does not negate the similarities, they are the same.

Leadbeader tells how a Tulpa like entity can exist for years feeding off of the thoughts and sacrifices of a village. It sounds eerily familiar to many of the artificial structures in our modern world:

By means of whatever nourishment they can obtain from the offerings, and still more by the vitality they draw from their devotees, they may continue to prolong their existence for many years, or even centuries, retaining sufficient strength to perform occasional phenomena of a mild type in order to stimulate the faith and zeal of their followers, and invariably making themselves unpleasant in some way or other if the accustomed sacrifices are neglected. For example, it is asserted that in one Indian village the inhabitants have found that whenever for any reason the local deity is not provided with his or her regular meals, spontaneous fires began to break out with alarming frequency among the cottages, sometimes three or four simultaneously, in cases where they declare it is impossible to suspect human agency.[16]

One can easily substitute corporation, institution, military-industrial complex, the market, socialism, communism, capitalism, money, science and scientific thinking for the deity in India which Leadbeater talks about. The fact is that any artificial construct/entity we create has the power to become a Tulpa, or a demonic like entity.

Humankind's history is full of false gods and seems to be growing. Frank Manuel in his analysis of the 18th century felt it was 'Les progress'[17] that had taken on God-like status. David Hawkin feels that the new gods of the 21st century are "those to which our Western Hegemonic culture seems increasingly to grant intimacy."[18] Christopher Catherwood sees mass murderers' who inspire as false gods: "Hitler, Stalin, Milosevic, and now Osama Bin Laden are examples of false gods—human beings who inspire in the followers such fanaticism that they would die in the cause that their leader espoused."[19]

Idolatry
Idolatry is the worship of false gods. You don't have to get on

your hands and knees and pray to be considered worshipping a false god in our modern world. We are constantly worshipping false gods by thinking about them, and altering our behavior to suit them through seemingly innocuous things such as focusing on popular culture in dress and entertainment. We are arguably forced to participate and bow to such false gods such as science and technology because we cannot exist without them.

Jewish philosopher Rabbi Moses Maimonides noted that the Hebrew prophets had their strongest rebuke against idolatry: "[I]n examining the Law and the books of the Prophets, you will not find the expressions 'burning anger,' 'provocation.' Or 'jealous' applied to God except in reference to idolatry; and that none but the idolater called 'enemy'' 'adversary,' or 'hater of the Lord." [20]

The prophets understood the dangers of false gods and spoke of the necessity of having an exclusive relationship with God like a marriage. Isaiah said, "For your Maker is your husband, the Lord of hosts is his name."[21] God asked Hosea to take on an adulteress wife to symbolize Israel's unfaithfulness with the worship of idols.[22] Similarly God through Jeremiah condemns Israel for its infidelity: "Instead, as a faithless wife leaves her husband, so you have been faithless to me, O house of Israel, says the Lord."[23] Moshe Halbertal and Avishai Margalit in their book *Idolatry* note that the imagery of marriage was meant to emphasize the exclusivity of Israel's relationship with God.[24]

Idolatry takes us farther away from our true nature and God. Jeremiah tells us that it is God that holds life and meaning and not false idols: "[F]or my people have committed two evils: they have forsaken me, the fountain of living water, and dug out cisterns for themselves, cracked cisterns that can hold no water."[25] Our world is filled with cracked cisterns that are lifeless and un-fulfilling.

Maimonides said that, "Idolatry is founded on the idea that a particular form represents the agent between God and His creatures."[26] The problem with this false identification according to Maimonides is that it evolves and eventually people accept it (form, rites) as salvation and stop believing in God: "By transferring that prerogative to other beings, they cause the people, who only notice the rites, without comprehending their meaning or the true character of the being worshipped, to renounce their belief in the existence of God."[27]

It is that substitution for God, whether it be an image, an idea, something that makes us feel good or important, or an association that leads us to idolatry. Instead of looking to God to find happiness, identity, meaning and purpose we look to an object, or an idea or a process. Money, what some call the root of evil is a form of idolatry.[28] Our modern world has created a myriad of possibilities and temptations for idolatry. Idolatry can form around faddish clothing, new technologies, famous people (Hollywood, sports, political, business, music, etc.), lifestyles, fine foods or wine, cults and their leaders, musical lyrics, movies, behavior, ideas, actions and more.

We need to look at idolatry with fresh eyes and not see God speaking to the Israelites as a jealous God[29] who demands monogamy but one who does so because of love and the desire to help us be liberated. God understands how the universe works and does not want us to get entangled with false gods and idols.

When we focus on an idol with a lot of thought we give it much power.

Artificial Structures

Whenever we create an organization, institution, a religion or structure of any sorts we are creating an entity. Inevitably an entity develops independent to the underlying principle, cause or purpose for what it was designed for, even if it is for the good. As with our thoughts, crowds and tulpas, these entities take on lives of their own and change from their original intent as the seed thoughts applied to them begin to mutate.

The hypocrisy of institutions is demonstrated very vividly with the Roman Catholic Church which has consistently through out its history defied and run roughshod over the teachings of Christ. The Crusades, the Inquisition, the selling of papal indulgences and most recently the pedophile scandal show the hypocrisy of the church and its disconnect from Jesus. The actions of Pope Benedict XVI to promote Cardinal Law of greater Boston Diocese after he resigned because of the pedophilia scandal shows that he was more concerned about the institution of the church, rather than Jesus. The church like the mob or crowd Le Bon describes, has taken on a purpose that flies in the face of what Jesus taught.

The prophets understood the power of institutions and organizations. No prophet ever advocated creating a religion or any other institution. Prophets taught us how to live, how to worship, to seek justice and most of all to have compassion and mercy for each other.

God through Amos tells us that he is not interested in sacrifices, festivals and trappings of organized religion. God wants fairness, equality; God wants JUSTICE:

> I hate, I despise your festivals, and I take no delight in your solemn assemblies. Even though you offer me your burnt-offerings and grain-offerings, I will not accept them; and the offerings of well-being of your fatted animals I will not look upon. Take away from me the noise of your songs; I will not listen to the melody of your harps. But let justice roll down like waters, and righteousness like an ever-flowing stream.[30]

Jesus never advocated forming a religion or any other institution. He even questioned the institution of the family and said that his brothers and sisters were those that did God's will:

> Then his mother and his brothers came; and standing outside,

they sent to him and called him. A crowd was sitting around him; and they said to him, 'Your mother and your brothers and sisters- are outside, asking for you.' And he replied, 'Who are my mother and my brothers?' And looking at those who sat around him, he said, 'Here are my mother and my brothers! Whoever does the will of God is my brother and sister and mother."[31]

Krishnamurti understood how religious institutions could subvert us from the truth. In his youth he was discovered in India by clairvoyant C. W. Leadbeater who upon seeing his aura, not a speck of selfishness in it, heralded him to be great teacher. In 1911 the Theosophical Society with great hope formed a religious institution, the Order of the Star around him. As he matured Krishnamurti came to understand that religious institutions like the Order of the Star diverted people from the path of God and Truth. In 1929 he dissolved the organization.

He began his dissolution speech with a story:

You may remember the story of how the devil and a friend of his were walking down the street, when they saw ahead of them a man stoop down and pick up something from the ground, look at it, and put it away in his pocket. The friend said to the devil, 'What did that man pick up?' 'He picked up a piece of Truth,' said the devil. 'That is a very bad business for you, then,' said his friend. 'Oh, not at all,' the devil replied, 'I am going to let him organize it.'[32]

Given the historical record of religious institutions—religious wars (crusades, jihad), violence, persecution of women/pagans/other religions, ethnic cleansing, pedophilia—one must wonder.

Krishnamurti went on to detail how religious institutions encumber us in the path to find the Truth and God:

Truth is a pathless land. Man cannot come to it through any organization, through any creed, through any dogma, priest or ritual, nor through any philosophical knowledge or psychological technique.

He has to find it through the mirror of relationship, through the understanding of the contents of his own mind, through observation, and not through intellectual analysis or introspective dissection. Man has built in himself images as a sense of security—religious, political, personal. These manifest as symbols, ideas, beliefs. The burden of these dominates man's thinking, relationships and his daily life. These are the causes of our problems for they divide man from man.[33]

The way we clamor for greater meaning and purpose, or what Le Bon calls religious sentiment, when we gather collectively shows that at our core we long for something more. Deep down we desire to part of something beyond, to have a greater purpose. Unfortunately, we have trapped ourselves in the physical plane and filled our mind with delusions and idols. There our vulnerability for self sacrifice and longing for the divine has been exploited in a myriad of causes and purposes other than God.

Our Collective Un-Conscious
Swiss psychologist Carl Jung tells us that we are made up of a primordial soup of the collective history of humanity. He believed that each of us has a piece of a universal 'collective unconscious' within us, what he called 'archetypes':

A more or less superficial layer of the unconscious is undoubtedly personal. I call it the personal unconscious. But this personal unconscious rests upon a deeper level, which does not derive from personal experience and is not a personal acquisition but is inborn. This deeper layer I call the collective unconscious. I have chosen the term 'collective unconscious' because this part of the unconscious is not individual but universal; in contrast the personal psyche it has contents and modes of behavior that are more less the same everywhere and in all individuals. It is, in other words identical in all men and thus constitutes a common psychic substrate of a super personal nature which is present in everyone one of us.[34]

He goes on to describe archetypes:

[S]o far as the collective unconscious contents are concerned we are dealing with archaic or—I would say—primordial types, that is, with universal images that have existed since the remotest times...The archetype is essentially an unconscious content that is altered by becoming conscious and by being perceived, an it takes its colour from individual consciousness in which it happens to appear.[35]

What Jung is describing sounds very similar to a thought form, or samskara. The 'primordial types' he talks about are seed thoughts planted long ago that we have built upon. What we think about reverberates in history.

Chapter 7

The Market (mob) as God

The market, unfettered global capitalism, unbridled trade and the commoditization of everything from material goods, to ideas, to people, rules the world. The market is the mob, a false god that has morphed and grown to immense power. Injustice, inequality, exploitation of the poor and defenseless, misery and abuse are its hallmarks. It preaches and demands violence. It indulges and rewards its devotees with material objects (talismans) and pleasures in the physical plane, sucking them ever more into the delusion of materiality. The market is the antithesis of all that God teaches and calls for, yet too many the market is God.

If we are to liberate ourselves we need to understand that the market is that dominate false god in the world today and that it manipulates us and has us worship it. See the market for what it is.

The Market is the New Religion

Theologian Harvey Cox tells us that in our modern world many believe that the market is God and that like God it is omnipotent, omniscient and omnipresent. If we do not believe that the market is God it is because we lack faith; "At the apex of any theological system, of course, is its doctrine of God. In the new theology this celestial pinnacle is occupied by The Market."[1]

Cox goes on to tell us that the false god of the market has evolved to the point where it accepts no other false gods or idols. In other words it has won out over competing idols, or converted them.[2]

Economic historian Bradford Long implicitly agrees with Cox when speaking of the triumph of economics in the twentieth century, whether you call it capitalism or communism. He feels that for the first time in history economics became the driving force in all that we do.[3]

Karl Polanyi says that society has been deluded into thinking that the market is a *'utopia'* that can resolve all its problems by giving everything a price. According to Polanyi the market has commoditized humanity and in doing so, "human society had become an accessory of the economic system."[4]

Many worship the market the same way others worship God. Holding vigil over the stock market has become the new national past time for many of us. Today instead of ringing of the church bells every 15 minutes 24/7 as was done in medieval times, the media barrages us continually with updates on the stock market, company earnings, economic statistics and other forms of financial news. Ticket sales (statistics) of movies have replaced reviews by critics as a barometer of the quality of a movie. Everything is judged by how it will affect the market. If the market goes up it is good, if it goes down then the news, or the event, is bad. The market has become like the primitive God by which the judgment of what is good or bad is determined by how it affects the crops.

The problem is that the false god of the market continues to grow and grow holding ever greater control over us. Even a disaster may well provide another opportunity to dominate us even more.

Dogma, Belief and the Mysticism

The market has a complex cosmology with various doctrines, paths to salvation and belief systems. David Loy says that, "[O]ur present economic system should also be understood as our religion, because it has come to fulfill a religious function for us. The discipline of economics is less a science than the theology of that religion, and its god, the Market, has become a vicious circle of ever-increasing production and consumption by pretending to offer a secular salvation."[5]

Economists, traders, portfolio managers and market pundits are the high priests of the market. They are given exalted status and cloaked with religious symbolism, called 'guru', 'oracle', 'sage' and the like. Financial market cable TV bombards us with their prognostications.

The core belief of the market, upon which all else flows is from, is based upon the rule of 'self-interest' which is in total opposition to God's call to love and to 'do unto others'. It was the capitalist's sage, Adam Smith whom elucidated this universal falsehood; "It is not from the benevolence of the butcher, the brewer, or the baker, that we expect our dinner, but from their regard of their own interest."[6]

The pursuit of self interest drags us deeper into the physical plane. It advocates total disregard for anyone but oneself. It turns love from a heart felt emotion of self sacrifice and giving to a mutually beneficial relationship, if love is possible at all for devotees of the market. From marrying for money, to buying a mail order bride to more temporal relationships, the market sees everything in dollars and cents.

Our first steps, our seed thoughts are critical, they are our foundation. When we plant a seed of our own self interest first and foremost we begin a downward spiral, both with our life in the physical plane and with our spiritual evolution. Self interest and its progeny of greed, materialism and its trappings of wealth and attachment, callous disregard for others, selfishness, hate and violence diminish and deplete our soul.

The mysticism of the market is seen through the power of the 'invisible hand.' It is the force and power of the 'invisible hand' that makes the market omnipotent, omniscient and omnipresent. According to Smith there is an 'invisible hand' that orchestrates all the machinations of the market—bringing sellers and buyers together, setting prices, advancing society, providing product choices and more. Devotees of the market believe that it is the 'invisible hand' that makes it the wonder that it is. Smith tells us that when we pursue our own 'self-interest', for example sell or make something, it is the 'invisible hand' that leads us for our own and societal benefit: "...he [sic] intends only his own gain, and he is in this, as in many other cases, led by an invisible hand to promote an end which was no part of his intention. Nor is it always the worse for society that it has no part of it. By pursuing his own Interest he frequently promotes that of the society more effectually than when he really intends to promote it." [7]

The 'invisible hand' guiding the market gives it mystery and God-like attributes. It also speaks about faith, that the market will save us and lead us. Smith notes how the market is the efficient arbiter of all: "As it is power of exchanging that gives occasion to the division of labor, so the extent of its division must always be limited by the extent of that power, or, in other words, by the extent of the market."[8]

The idea that an 'invisible hand' links participants speaks to group consciousness. Market participants are linked by one thought, 'self-interest.' The term 'invisible hand' is Smith's way to describe the link and the power of group consciousness to transcend the physical plane. The 'invisible hand' like the crowd Le Bon describes gives hope to the devotees of the market that they are part of something greater than themselves.

Belief in the 'invisible hand' gives absolution to market devotees. They believe as Smith says that, however they pursue their own self-interest it will benefit society. So when they trample the poor, desecrate the environment and proselytize with the products they are trying to sell, they are doing their false god's work.

Entrepreneurs tell us that they are developing and providing new products that will benefit society. They remind us that their companies are creating new jobs and in doing so helping society. This belief eases their conscience and justifies their actions no matter how cruel, inhuman or unjust they may be.

Capitalism preaches property rights, the power of material things, but do we have the right to parcel up and exploit the land, especially if we understand that all, in this case the land, is Brahman? Similarly the Abrahamic tradition tells us that, "The land shall not be sold in perpetuity, for the land is mine; with me you are but aliens and tenants."[9] So we must ask ourselves are we honoring the divine and Mother Earth as when we are misusing this sacred earth?

The ownership of land gives us a sense of power and the belief that it is ours to do with it as we like. The thought form of ownership also bears fruit of owning other things. Ched Myers tells us that the market has led to our loss of God's gifts and created a different and alternative world;

> Our modern economic system is a cruel parody of Sabbath economics. We have indeed created a situation of 'scarcity' in nature because of our relentless plundering of the land and its gifts. At the same time, capitalism has demonstrated incredible ingenuity and capacity to manufacture 'artificial Abundance' (think of the 30 kinds of cereal or toothpaste at your local supermarket). There is enough, but it is no longer a gift of creation. Rather, it is a marketed commodity, which by definition does not circulate equitably to everyone. Thus we have 'reengineered' the world: our refusal to limit our appetites has drained natural abundance, and our artificial abundance belongs only to the few. This is not ironic: It is idolatrous.[10]

The Temple

All religions have a temple, a place of worship that is sacred and holy and held in high esteem. The temple is a link to the divine. It is a place held in awe by its members and where they go when they need answers.

Central banks are the market's temple, most notably the

Federal Reserve of the United States of America. William Greider understood this when he wrote a book about the Federal Reserve (the Fed) and titled it: *Secrets of the Temple How the Federal Reserve Runs the Country*. Greider describes the Fed as follows:

> Like all conspiracy theories, the ones aimed at the Fed were confused attempts to confront larger mysteries of life, to explain the awesome powers that were shielded from the scrutiny of ordinary mortals yet seemed to govern their lives, like the temple incantations of ancient priests who interpreted divine messages and decreed the course of social destiny. In a twisted sense, belief in a grand conspiracy was an act of religious deference, an acknowledgement by people that someone or something held distant and unexplainable power over them....Somewhere, in a hidden place, there were mortal men who conspired to rule over all—to usurp the powers that belonged only to God.[11]

The language of the Fed—money supply, Fed Funds rate, velocity—and its secretive nature of closed door meetings and cryptic prognostications give it an air of mystery.

Whenever there is a crisis in the markets, as with the pestilence or calamity of old, everyone rushes to the temple for help. Each major crisis over the last few decades from the 1987 stock market collapse, the collapse of the real estate bubble beginning in 2007, or the Bear Stearns and Fannie Mae/Freddie Mac rescues in 2008 and the $700bln. bailout of banks, market participants immediately looked to the Fed for help. Help has come with soothing words and policy actions, often bailouts and easy money. It should be pointed out that bailouts fly in the face of self-regulation and the winnowing process preached by free market capitalists.[12]

James Livingston states that the Federal Reserve was created early in the twentieth century to protect the interests of a new ruling class he called the 'corporate-industrial business elite.' By establishing the Federal Reserve the wealthy upper classes were able to seize control of the USA, not by buying votes or rigging elections, "but by establishing and enforcing the cultural or ideological consensus within which public debate on banking problems and solutions took place."[13]

Deities

Corporations are the markets dominate deity-like entity. Unlike deities, or angels, that are munificent and help with our evolution, corporations are selfish, predatory and demonic in nature. Like a tulpa they are an artificial construct, a false god, an entity that ultimately takes on a life of its own. Corporate entities are created primarily to absolve shareholders of any legal liability (sin) should they break the law or damage something or someone and incur huge liabilities. In other words, they place their shareholders above the law.

Corporations began taking form in the seventeenth century with the advent of capitalism. Their power and stature was greatly increased in 1886 when the Supreme court ruling in St. Clara County vs. Southern Pacific Railroad invoked the 14th amendment and defined corporations as persons. Once given the legal status of personhood a corporation had 1st amendment rights and with it enormous power.

Throughout the twentieth century corporations would look to exploit, manipulate and gain favor at the expense of citizens around the world. Their fingers were constantly pressing government to curry favors—tax breaks, subsidies, rulings that limited competition, sweetheart deals and reduction of regulatory controls. The term 'corporate welfare' explains their gluttonous behavior.

By the end of the twentieth century corporate greed and power had grown so massive that Kevin Phillips said that the only comparable period was America's gilded age. He said that it was the mega merger of politics and the market that allowed money to drive policy that led to the loss of democracy.[14]

Not only were corporations squeezing the government for more, they were so greedy that they demanded the government cut back support for the poor and disenfranchised. Ralph Nader notes that while President Reagan decried fictitious welfare queens that were robbing the government it was President Clinton that put the poor out on the street by gutting welfare and turned a blind eye to corporate greed and abuse.[15]

As with all thought forms and false gods the power of corporations has continued to grow and expand. In 2001 the special status granted corporations under the 1993 NAFTA and chapter 11 ruling had given corporations status above countries.[16] Corporations could now sue countries if their markets were closed to them or placed what they felt were onerous regulations on them. Corporations have also begun enlisting the World Bank and the IMF to spread the gospel of free markets by forcing heavily indebted third world countries to accept privatization so that they can freely divvy up their assets.[17]

In 1886 the courts defied the constitution of the USA when they granted special privilege to corporations and made Americans second class citizens. The power of corporations has continued to grow to where they now hold sway over countries. What is next?

Age of Hyper-capitalism

At its core the market is the barbarians, the thundering herd of old, who raped and pillaged for gain, or what Le Bon called the crowd, the ultimate mob. The most recent manifestation is large marauding pools of capital that scour the world looking for plunder.

There have been a plethora of attacks and panics on the worlds' financial markets by speculators. In 1992 sensing vulnerability and an opportunity to pirate gain speculators mercilessly attacked (sold short) the British pound with reckless abandonment. Britain, as the UK Guardian reported was, "defenceless against international currency speculators"[18] and forced to depart from the European Exchange Rate Mechanism (ERM). In the months afterwards a host of European countries would be similarly attacked.

Gregory J. Millman in *The Vandals' Crown; How Rebel Currency Traders Overthrew the World's Central Banks* describes the enormous power of speculators; "Like the vandals who conquered decadent Rome, the currency traders sweep away economic powers that have lost the power to resist."[19] The Times of London noted how speculators have been able to do what OPEC could not do, fix prices: "Traders can hold the world economy to ransom because short-term demand and supply are inflexible, but also because they dominate dealings on oil markets... The tail wags the dog in many commodity markets, wildly exaggerating the ups and downs of demand and supply. In oil markets, the tail wags an elephant."[20]

The sacrilegious belief among punters and free marketeers is that they are the righteous force of the market's 'invisible hand' enforcing discipline on the weak and irresponsible. As Millman notes, "Like bounty hunters in the Old West, the traders enforce the economic law, not for love of the law; but for profit."[21]

The righteous force of the market is violence, whether it is the plunder of innocents by speculators, or outright war or genocide. Historian David Hackett Fischer says that over the last 800 years achieving price stability and market equilibrium has come with pain, hardship and death; and when equilibrium has been achieved it has been fleeting and not lasted long.[22]

Money a beast unto itself

Money is the handmaiden of the market and a powerful talisman. From its conception in the earliest days, it has been a force for power and control. It draws us more into the physical plane of materialism. Money creates attachment and separation from Source that ultimately leads to violence as we fight for power and possessions. It creates addictions—greed, possessiveness, desire.

Money has a consciousness attached to it, greed, injustice and materialism. When we touch it we come in contact with that consciousness. Handling money, and thinking about money as with all thought forms, gives it strength. It is a powerful talisman for darkness. Through out history money has led the charge against God's teachings of love and justice.

Money is an idol of the false god of the market. Jesus told us that "No one can serve two masters; for a slave will either hate the one and love the other, or be devoted to the one and despise the other. You cannot serve God and Money."[23] Humankind has always looked to money to replace God in our hearts. We think that money will make us happy, give us identity, gives us power and stature, ensure our survival and reduce our necessity to work.

Karl Marx saw that the power of money in the nascent beginnings of the industrial revolution:

Money degrades all the gods of mankind—and converts them into commodities. Money is the general, self-sufficient value of everything. Hence it has robbed the whole world, the human world as well as nature, of its proper worth. Money is the essence of man's labor and life, and this alien essence dominates him as he worships it....

The metallic existence of money is only the official sensuous expression of the very soul of money existing in all branches of production and in all operations of civil society....

The more the worker exerts himself, the more powerful becomes the alien objective world, which he fashions against himself, the poorer he and his inner world become, the less there is that belongs to him.[24]

The continued thought and desire (emotion) applied to money has given it even more power. The poor give it power by worrying(emotion) about not having enough to make ends meet. The market has created cataclysms of financial panics, bank runs and economic depressions that have led vast numbers of people to worry about money; all of this emotional worrying have given it even more power. Money has gained power and stature that would have been unimaginable not long ago and in doing so has borne ever more malicious and wicked fruit. Fear is a powerful force.

Money Morphs and Spawns Derivatives

Like all thought forms money has begun to mutate in hereto unimaginable ways. A host of new financial products, money surrogates, have been created since the 1970's that have redefined and expanded the concept of money. Some of these such as credit cards gave people the opportunity to spend beyond their means. Others such as money market funds, provided higher rates of interest than traditional bank savings deposits and were very liquid because checks could be written from one's account.

The most notable mutation of money was the creation of derivatives. Derivatives, as the name implies, are securities whose value and existence are derived from another security. For example, an option (derivative) would give you the right to buy or sell a stock at a particular price; futures allow one to invest in the price appreciation or deprecation of a bond or currency for a fraction of the cost of investing in the underlying security itself. Derivatives add to the casino like nature of the financial markets.

Derivatives almost non-existence prior to 1973 had a face value of $683 trillion by 2008 according to the Bank for International Settlements (BIS).[25]

Derivatives helped facilitate the flow of large pools of capital across international borders. This mobility of capital fostered the process of globalization. It also allowed for the development of a pool of large speculators from the trading desks of brokerages to hedge funds that could lay siege to countries and markets.

Derivatives, like all thought forms would continue to evolve and expand in un-imaginable ways. One of which was the creation of investment vehicles to exploit the poor. Financial deregulation in the USA in the 1980's began an exodus of banks from inner cities and the redlining (not giving loans) in less affluent parts of a city, or to people of color. Redlining deprived potential homebuyers the opportunity for mortgages, thereby increasing the likelihood that homes would be abandoned because they could not be sold, contributing to the overall decline in home values in red-lined neighborhoods.

The departure of banks from poorer neighborhoods created a vacuum and left a lot of people un-banked.[26] Estimates as to the number of un-banked in the late 1990's and years following was conservatively estimated at 9.5% to 20% of the American population.[27] Not having a checking or savings account creates a myriad of problems and makes people second class citizens. In a world dependent upon finance they were excluded. Excluded meant that opportunities, such as getting a mortgage to buy a home, or having some place other than a cookie jar for their savings were unavailable to them. It also meant relying on expensive cash checking outlets to pay bills.

The market for the un-banked, or fringe banking, would continue to grow in the 1990's into the hundred of billions of dollars[28] in size and develop new products such as auto title loans, pay day loans, tax refund anticipation loans, sub prime mortgage loans, rent to own stores and more. Ingenious methods were developed to circumvent state usury laws so that exorbitant interest rates, which could reach into the hundreds of percentages annually, could be charged. Often loans were made knowing that the borrower would have to take out another loan to make payment on the original loan and eventually be caught in the web of a large debt overhang.[29]

Brainwashed and Branded

The Bhagavad-Gita tells us that we achieve liberation by entering the deep trance state of Samadhi and breaking free of the miseries created by the physical plane and sense objects.[30] Conversely the market looks to entice us into the material world of sense objects and pleasures. It rewards the rich with pleasures and plunders. It offers a plethora of talismans, sense objects of dead nature, arranged in a hierarchy of value based upon price. As the false god of the market has grown and metamorphosed it has manipulated and demanded more of its devotees. We have progressively been induced to buy more products and services and have increasingly come to identify with and worship them.

William Leach in *Land of Desire* notes that a distinct culture of consumerism not associated with the traditional family, or American politics or even religious institutions was introduced by merchants after the Civil War. It was a culture based upon money and the purchase of goods: "The cardinal feature of this culture were acquisition and consumption as the means of achieving happiness, the culture of the new; the democratization of desire; and money value as the predominant measure of all value in society."[31]

By the middle of the twentieth century selling had burgeoned beyond the extravagant displays of merchandise used by the Wanamaker's and Macys to induce people to buy goods as described by Leach. Corporations and emerging advertising firms were resorting to highly sophisticated means to prey on people's fears and insecurities. Vance Packard in *The Hidden Persuaders* commented on how,

> The use of mass psychoanalysis to guide campaigns of persuasion has become the basis of a multi-million dollar industry. Professional persuaders have sized upon it in their groping for more effective ways to sell us their wares— whether products, ideas, attitudes, candidates, goals or states

of mind...

Certain of the probers, for example, are systematically feeling out our hidden weaknesses and frailties in the hope that they can efficiently influence our behavior.[32]

Packard noted that an ad executive in a trade magazine talked about how the cosmetic industry grew to become over a billion dollar industry by selling hope and making women feel anxious and critical of their appearance.[33]

He asked, "what are the ethics of businesses that shape campaigns designed to thrive on these weaknesses that they have diagnosed?"[34]

By the end of the twentieth century the market and its fruit of selling had mutated and were bearing even more nefarious fruit. Naomi Klein says that corporations, through their marketing efforts, image making and logos, had begun branding consumers as if they were cattle in an effort to have them identify with and find meaning in their products.[35] Kalle Lasn of Adbusters says that "America is no longer a country. It's a million dollar brand...Our mass media dispenses a kind of Huxleyan 'soma'."[36] He says that we are constantly fed 'infotoxins' and 'media viruses' that turn us into 'Manchurian consumers' becoming like Sergeant Raymond Shaw in the *Manchurian Candidate*. So many of us are very careful about what we eat and yet we ignore the 'infotoxins' that our consciousness is constantly absorbing that ultimately affects our health and spiritual well being.

The Beast Grows and Needs more food

America like much of the world has had its natural treasures plundered so that there is not much left. Like Genghis Khan's thundering herd the market and its devotees have devoured all like a plague of locust. Tax breaks for the rich, corporations and special interests along with un-bridled military spending and massive bailouts have turned the USA's finances deep into the red. Similarly much of Europe and the developing world is mired in debt and/or had their natural resources plundered through privatization. Beginning with Reagonomics, what President G.H. Bush called "voodoo economics", America went from being the world's largest creditor to becoming the largest debtor in the world.

We have been running up trade and current accounts deficits for years that even the most optimistic such as ex-Fed Head Alan Greenspan[37] thought would have precipitated a massive adjustment. American citizens are similarly are drowning in debt. The market has also stripped America and the world of much of its other assets through privatization.

Although the cupboards are bare the market still wants to be fed, to occupy and dominate our thinking even more. It is a false god with an insatiable appetite that gets only larger. Like Machiavelli's *The Prince* it does not care whether we love it, or fear it. It does not care whether you call it capitalism, communism, socialism or laissez-faire as long as you focus and dedicate yourself to the worship and sacrifice of money and its talismans. A strong emotional fear or terror, or a strong passion for things like a drug addict's longing for drugs elicits a strong emotional thought. The false god of the market craves such thoughts as these.

Where will the market turn having stripped the cupboards? Economic calamity and war are some of the possibilities. Whatever the choice a world dedicated to the market is sure to see more misery, injustice and war, all of which would feed it with strong emotional thoughts.

Chapter 8

An Abundance of Snares

The market is not the only beast that we face. A multitude of false gods and negative thought forms hold sway over us. As we have sunk deeper into the physical plane our attention has been increasingly focused away from God and towards them.

These false gods and negative thought forms have grown so much that it is impossible to escape them. They have manipulated the physical plane for their own ends and entangled us in their webs. To live, to eat and to survive we must participate in their charade and in doing so we give them strength. They incessantly bear more fruit and each successive crop becomes more malicious, more controlling. Each day we become more vulnerable and dependent.

Much of what we hold dear and cling to for spiritual, emotional and physical safety and security from religious institutions to religious beliefs to the government are not protecting us. Instead they are like an anchor that plunges us deeper into the abyss of darkness. If we are to break free it is important to understand what is blocking our path to liberation.

Some may feel that what I am pointing out is preposterous; others may feel that I have not gone far enough. To understand the morass we are in, it is necessary to look at the world through the eyes of a prophet to see all that ails us. Deceptions and snares abound, here are a few.

Marriage of Greed and the Gospel
The voice of the prophets and scripture have long been subverted, the gospel of today is 'the market', our dominant false god. Many believe in a new gospel of prosperity, that marries the

greed of capitalism with scripture. Called 'prosperity theology', or 'prosperity doctrine', or 'prosperity gospel', it says that it is not only okay to be rich, but a sign of God's favor.

Services are often held in large mega churches that can hold thousands if not tens of thousands. These mega churches resemble malls and offer a plethora of visceral experiences to the church goers. They are run as businesses rather than community with its vestry and participation by all its members.

The gospel of prosperity preachers is neither totally demonic nor totally divine, but rather it is gray. They advocate faith and belief in God, but at the same time they advocate the accumulation of wealth and material possessions. What is so threatening about many of these prosperity preachers is that they are implicitly advocating techniques that work with thought forms. Instead of using techniques and practices taught by sages to help us better connect with God and make the world better they focus their efforts on wealth and material objects.

Dr. John Hagee, senior pastor of the Cornerstone church in San Antonio in *The Seven Secrets*, writes how important money is because "five times more is said about money than prayer....every spending decision is a spiritual decision."[1] Hagee feels our stewardship of money is so important because it affects others and that on judgment day God will judge us by how we managed our wealth. He makes a sharp distinction between 'self-interest' that is good versus 'selfishness' that is bad. Like many prosperity preachers Hagee has a very distinctive view of heaven: "The eternal kingdom has twelve gates of solid pearl with streets of fine-spun gold. Mansions left and right are designed by the architect of the ages for God's children. If that's poverty, I say bring it on!"[2]

By providing a vision of heaven being made of gold, pearls and other objects of wealth Hagee is saying that is what we should aspire for. Hagee is asking us to 'visualize' heaven as being a gaudy palace filled with the entrapments of the material world. Remember that we become what we think and if we are told that heaven is about material wealth then we will look to bring material wealth into our life.

Joel Osteen in his *Your Best Life Now* says that it is time to reap God's rewards here and now in the physical plane. In an eerie way he is recommending that we create a thought form focused primarily on materialism and personal gain to guide our lives: "It is time to enlarge your vision. To live your life now, you must start looking at life through eyes of faith, seeing yourself rising to new levels. See your business taking off. See your marriage restored. See your family prospering. You must conceive it and believe it to be possible if you ever hope to experience it."[3]

To achieve wealth and success Osteen preaches thinking positively and understanding how your mind works. Although he does not call it the law of attraction he does say; "like a magnet, we draw in what we are constantly thinking about."[4] He understands the power of the mind to transform our lives and preaches about it. One wishes that he focus more on God and the spiritual and less on physical matters.

One of the criticisms of prosperity theology is that it ignores the gospel of social responsibility preached by Jesus and the prophets. When Osteen advocates "expect preferential treatment," [5] and "the favor of God"[6] he is planting a seed thought of superiority and inequality that reinforce societal inequalities.

New Age False Prophets
Similarly many in the New Age world embrace the catechism of greed thinking that our purpose in life is to increase our wealth and to enjoy its spoils. They talk about the Law of Abundance and how we can have all the wealth, things and pleasures we want. To achieve this abundance they tell us that we need only focus our thoughts on acquiring material possessions. These New Age preachers like their more traditional Christian

brethren, talk about how our thoughts affect our lives and livelihood.

Rhonda Byrne in the highly popular *The Secret* says that all the money in the world is out there for us and all we need to do is ask for it. She basically advocates creating a thought form of pulling in money and wealth into your life, what she calls abundance:

The only reason any person does not have enough money is because they are blocking money from coming to them with their thoughts. Every negative thought, feeling, or emotion is blocking your good from coming to you, and that includes money. It is not that the money is being kept from you by the Universe, because the money you require exists right now in the invisible. If you do not have enough, it is because you are stopping the flow of money coming to you, and you are doing that with your thoughts. You must tip the balance of your thoughts from lack-of-money to more-than-enough-money. Think more thoughts of abundance than of lack and you have tipped the balance.[7]

One of the secrets Byrne reveals to her readers is that the wealthy, who she thinks we should admire, are those that obsess about money and become so focused on wealth that let nothing else enter their mind:

People who have drawn wealth into their lives used The Secret, whether consciously or unconsciously. They think thoughts of abundance and wealth, and they do not allow any contradictory thoughts to take root in their minds. Their predominant thoughts are of wealth. The only know wealth, and nothing else exists in their minds. Whether they are aware of it or not, their predominant thoughts of wealth are what brought wealth to them. It is the law of attraction in action.[8]

Talk about becoming what you think about!

By embracing and aspiring wealth, whether we are following the gospel of prosperity or the law of abundance, we are directly advocating for 'the market' and its precepts of self-interest and violence. Our participation in the market system gives it strength, by striving for wealth we take a lead role in its advance and become one of its high priests. Ched Meyers tells us that "disparities in wealth and power are not 'natural' but the result of human sin."[9] The prophets understood that wealth was achieved by oppression of the poor and disadvantaged, Jeremiah said: "Woe to him who builds his house by unrighteousness, and his upper rooms by injustice; who makes his neighbors work for nothing, and does not give them their wages."[10]

Ron Sider tells us that "God actually works in history to cast down some rich powerful people."[11] Jesus preached good news to the poor and condemned the rich and wealthy: "Blessed are you who are poor, for yours is the kingdom of God. Blessed are you who are poor, for yours is the kingdom of God....But woe to you who are rich, for you have received your consolation."[12]

By focusing on money and becoming wealthy you are in direct opposition to God's call to love and are diminishing your soul. Jesus told us that wealth is an encumbrance to our spiritual advance when he said "it is easier for a camel to go through the eye of a needle than for someone who is rich to enter the kingdom of God."[13] Earlier in discussing thought forms psychologist Frances Vaughan noted how using positive thinking to obtain ego needs such wealth can "imprison our soul."[14]

Wealth builds up bad karma that may require several more additional life times to work out. When you make the possession of money and wealth as a goal in your life, you create and strengthen samskaras of greed and selfishness. The entrapments of wealth and luxury that you acquire are negative talismans that cast a dark shadow over you and suck you deeper into negativity.

The Delusion of Democracy

With government we have put our faith in an institution over God and because of this we have paid a heavy price. As with all artificial constructs, government has transmogrified and become an unfamiliar entity that serves the interest of a handful of participants rather than the voters and the greater good.

The problem with government goes well beyond the influence of special interest and money, and rampant corruption as seen with political favoritism. The fact is that government is increasingly becoming the exact opposite of democracy—a tool of injustice and persecution. Whenever we give power to someone or something other than God we create or add to a thought form of superiority and control. Ultimately, as with all thought forms not focused on the divine, power manifests itself in diabolical ways.

In 2003 President Bush took the United States of America to war when he invaded Iraq. His administration lied about the rationale for attacking Iraq. Weapons of Mass Destruction were never found in Iraq as he claimed and Saddam Hussein did not have extensive ties with Al Qaeda.[15] Five years after the invasion Vice President Cheney was oblivious to the fact that 2/3rds of Americans were against the war.[16] A lot of reasons have been given as to why the USA invaded Iraq from oil, to President Bush avenging his father, to appeasing the Neocons. What is not in doubt is that the war is highly un-popular and the government defied public will.

The Bush administration used fear to excite the American public into wanting to invade Iraq. Fear has long been used by various administrations to control and manipulate the American public, particularly for war and military spending. After the 9-11 terrorist acts America was traumatized and feared another attack. The Bush administration fanned the flames by attacking the American public through measures like the Patriot Act and having the NSA spy on American citizens. Spying on Americans, harassing them and intimidation did not make America safe but added to the thought form of fear.

Moshe and Avishai in *Idolatry* say fear is a form of idolatry that has long been used to control the masses. Citing Maimonides and Spinoza they note that,

Idolatry is the manipulation of the imagination for controlling the masses by means of an image of the world built upon meaningless promises and threats. These promises and threats, in the name of gods and demons that are products of the imagination, constitute a dangerous substitution for a causal understanding of the world...

The order of things, according to this view, is as follows: the uncertainty in our world as human beings is a constant source of fear and anxiety. Increasing or decreasing this fear constitutes an outstanding means of control in the hands of an idolatrous leader. The fear itself increases both tendency to imagine and the strength of the imagination. The masses are dependent upon their imagination because they lack the critical intellectual faculty required to harness the imagination and attain appropriate causal knowledge.[17]

Fear is a very powerful thought form because it has such a high emotional component. Ultimately fear reaps what it sows—violence against the fearful. Unfortunately all the thoughts of fear generated by a variety of administrations and the military is going to come back to America.

Waging War on Americans

It is important to reflect upon the justice system in the USA. America by far has more prisoners, both in absolute numbers and on a percentage basis than any other country in the world. One out every one hundred Americans is in jail.[18] A disproportionate number of those are black men (1 out of 15) and Hispanic men (1 out of 36).[19] Why?

Many feel that the Federal Justice system like much of Washington is run by political elite. Ward Churchill has documented[20] the war that has been waged against Americans by the Justice Department. According to Churchill the FBI's, "raison d'etre is always been the implementation of what the Bureau formally designated from the mid-1950's through the early 1970's as 'COINTELPRO's' (COunterINTELLigence PROgrams) designed to 'disrupt and destabilize,' 'cripple,' 'destroy,' or otherwise 'neutralize' dissident individuals and political groupings in the United States, a process denounced by congressional investigators as being 'a sophisticated vigilante operation."[21]

Churchill details the covert and illegal actions taken by the FBI to disrupt groups within the USA: illegal spying, prosecutorial harassment, spreading rumors through hideous means such as 'snitch jackets,'[22] planting information/evidence, violence, encouraging warfare between groups and taking actions that helped lead to the death of American citizens they deemed a risk to security.

In the 1970's the Church Committee, the United States Senate Select Committee to Study Governmental Operations with Respect to Intelligence Activities, investigated how USA intelligence agencies conducted illegal operations against Americans. In their second report they called the government's use of intelligence "A New Form of Governmental Power to Impair Citizens' Rights."[23] In its final report it exposed the ugly reality of COINTELPRO:

COINTELPRO began in 1956,; it ended in 1971 with the threat of public exposure. In the intervening 15 years, The Bureau conducted a sophisticated vigilante operation....The unexpressed major premise of much of COINTELPRO is that the Bureau has a role in maintaining the existing social order...Under the COINTELPRO programs, the arsenal of techniques used against foreign espionage agents was transferred to domestic enemies. As William C. Sullivan, former Assistant to the Director, put it; 'This is a rough, tough, dirty business, and dangerous. It was dangerous at times. No holds were barred....We have used [these techniques] against Soviet agents. They have used [them] against us.'... Mr. Sullivan's description — rough, tough, and dirty — is accurate. In the course of COINTELPRO's fifteen-year history, a number of individual actions may have violated specific criminal statutes; 24 a number of individual actions involved risk of serious bodily injury or death to the targets (at least four assaults were reported as "results.)[24]

The artificial construct of the Justice dept like all entities looks to manipulate and control and continually bears fruit. Since the COINTELPRO era it has grown progressively stronger. President Bush used the terrorist attacks of 9-11 as an opportunity to initiate the War on Terror and attack Muslims in America and around the world. Within two years of 9-11 over 6,400 individuals, primarily Muslims, were targeted for having committed terrorism; very few of them were charged or convicted of terrorism.[25] The Washington Post found that those charged with terrorism and convicted, more often than not, were convicted of other, often minor charges.[26] The arrests were used as a ruse to intimidate and harass ten of thousands of Muslim Americans. For example, 150 Muslim families that had donated to the Help the Needy charity in Syracuse, NY, whose founder Dr. Rafil Dhafir was called a terrorist but never charged with terrorism,[27] were asked inappropriate questions in an intimidating fashion by the FBI.[28]

The attack on America by the Bush administration was not confined to persecuting Muslims in America. In 2006 a scandal broke out when eight US attorneys were fired because they refused to go after political targets of the Bush administration. Senator Patrick Leahy who led the Senate Judiciary Committee investigating the firings of US Attorneys said that the "United States Attorneys may serve at the pleasure of the president, but justice does not serve at the pleasure of the president, or any president."[29]

Servants not Rulers

Power by its very nature implies that someone or something has control or sway over others. When we give power to someone, or something like the government, we create and or add to a thought form of control and inequality. The prophets understood that power whether it be the king's, the priest's or the wealthy corrupts and brings injustice. Hosea said, "Hear this, O priests! Give heed, O house of Israel! Listen, O house of the king! For the judgment pertains to you."[30]

Jesus similarly echoed that power brought abuse and led to hypocrisy with his woes to religious leaders at the time.[31] In his parables on several occasion he said that he who is first will be last and he who is last will be first,[32] implying that those in power will be the last to get to heaven. Jesus tells us that leadership means service both to the cause and to the people: "You know that among the Gentiles those whom they recognize as their rulers' lord it over them and their great ones are tyrants over them. But it is not so among you; but whoever wishes to become great among you must be your servant, and whoever wishes to be first among you must be slave of all. For the Son of Man came not to be served but to serve, and to give his life as a ransom for many."[33]

The Native American prophet the Peacemaker understood the corruptive nature of power. In the Great Law of Peace he lists a variety of moral and spiritual qualities that leaders should possess. The idea that elected politicians should be servants of the people was born with the Great Law of Peace. The

Peacemaker instructed the Haudenosaunee that their chiefs be very thick skinned in order tolerate criticism and be totally committed to serving:

> The Lords of the Confederacy of the Five Nations shall be mentors of the people for all time. The thickness of their skin shall be seven spans— which is to say that they shall be proof against anger, offensive actions and criticism. Their hearts shall be full of peace and good will and their minds filled with a yearning for the welfare of the people of the Confederacy. With endless patience they shall carry out their duty and their firmness shall be tempered with tenderness for their people. Neither anger nor fury shall find lodgment in their minds and all their words and actions shall be marked by calm deliberation.[34]

While the constitution of the United States of America may be modeled after the Great Law Peace, elected officials and civil servants rarely, if ever, embody the model of behavior espoused by the Peacemaker. Has a politician ever washed anyone's feet the way Jesus washed his disciples feet?[35] The fact is that politicians are too busy raising money from special interests for re-election.

Theologian James Alison sums up religious and political leadership well: "The narrative of the Passion, with its central Icon of God revealed in an innocent human being handed over to torture and death by the religious and political authorities of the time, has a power which tends to put into question all political and religious authority."[36]

Gurus, Priests, Imams, Monks, Teachers
Power and corruption also affects our daily life and contacts. Teachers, religious leaders, community leaders hold sway over us. Swami Vivekananda tells us that our guru or religious leader

should be selfless and more concerned about teaching truth instead of getting credit for what they did:

All the Teachers of humanity are unselfish. Suppose Jesus of Nazareth was teaching and a man came and told him: 'What you teach is beautiful. I believe that is it the way to perfection and I am ready to follow it; but I do not care to worship you as the only-begotten Son of God.' What would be the answer of Jesus of Nazareth? Very well, brother, follow the ideal and advance in your own way. I do not care whether you give me the credit for the teaching or not. I am not the shopkeeper: I do not trade in religion. I only teach truth and truth is nobody's property. Nobody can patent truth. Truth is God himself. Go forward.[37]

The unfortunate thing is that too many of our religious leaders, self-help experts and gurus are concerned about power and getting credit. Some even live extravagant lifestyles in contrast to the path of renunciation taught by all the prophets and sages. They can all talk the talk, but they cannot walk the walk of renunciation and poverty.

Krishnamurti similarly saw the destructive power of religious leaders: "All authority of any kind, especially in the field of thought and understanding, is the most destructive, evil thing. Leaders destroy the followers and followers destroy the leaders. You have to be your own teacher and your own disciple. You have to question everything that man has accepted as valuable, as necessary."[38]

While it is natural to feel love and gratitude towards a teacher who has helped you, it is better to give thanks by helping someone else. I disagree that we should worship our gurus, religious leaders and spirit guides. While many may disagree, devotion (bhakti) should be to God and not our teacher. Understand that when you worship someone other than God you are creating a thought form of superiority. Once you create that thought form it is not going to differentiate between idolizing a maniacal despot from a loving and caring guru. Your thoughts and actions are adding to the thought forms of superiority, inequality and imbalance.

The problem with all this abuse of power, aggrandized egos and worship of the powerful, is that it has created a world of idols, belief in entitlements by the powerful, imbalances and differences between people. While the thought form of power and special status is a very old one, it has transmogrified and spread dramatically over the last century. As a consequence whole industries such as Hollywood have been created around the worship of screen 'idols', who are basically nothing more that shape shifters. Hollywooditis has spread to musicians, sports figures and hosts of others. Remember they are called 'idols' for a reason!

We must endeavor to put ourselves last and not first in all that we do. We must endeavor to be servants not rulers.

Talismans cover Mother Earth

We have blanketed Mother with our negative thoughts and dead objects that ooze polluting and violent consciousness. This consciousness is absorbed by us and prevents Mother Earth from conveying her positive essences to us. Garbage and pollutants can be geographically contained to a particular area, but this does not stop them from affecting us. Toxic sites, whether they are dumps or factories spewing pollutants, are like a cancerous cell on Mother Earth. It diminishes her ability to emanate beneficial essences and to heal herself. Spirit lines, lines of consciousness, passing through toxic areas pick up that negative consciousness and transport it along their path. You may be a hundreds of miles away from a nuclear waste dump, but if a spirit line that passes through the dump passes through your bed room you are in intimate contact with it.

We need to think about how we spread and create detrimental talismans in our daily lives. The chemicals we put down to treat our lawn not only enter the water table but their consciousness lingers in our yard long after they are physically gone. That bright manicured lawn is emanating death and destruction. Is it any wonder childhood cancer has been on the rise in recent decades? It is the same for chemicals and toxins we store in our garage or in our medicine cabinet.

Then there are the talismans of greed and consumption. The trappings of materialism, greed and avarice carry with them the violence and abuse that led to their creation. The larger the home, the more expensive the car, the more sophisticated and expensive the electronics the more negative will be the consciousness associated with them. Think of it like the food chain—the tuna contains as much mercury as all the smaller fish it has eaten.

Our homes are full of talismans of violence. Guns in particular draw and encourage violence and killing. Some people are deluded into thinking that owning a gun makes one safer—it does not. Violence attracts violence and a talisman of violence does not care whether you kill in self defense, are killed, commit suicide or gush violent thoughts. Owning a gun whether legally or not, attracts and promotes violence.

We are storing up violence with our military bases, arsenals, nuclear weapons, weapons manufacturing industries and missile defenses. As was noted earlier the atom bomb represents the culmination of eons of humankind's thoughts devoted to killing the enemy. All weapons similarly carry a similar thought form of killing that has been built over a long time; As such large and destructive weapons are powerful talisman's that look to reap violence.

Amos tells us that storing up weapons is wrong and will eventually bring calamity and violence to those who possess them: "They do not know how to do right, says the Lord, those who store up violence and robbery in their strongholds.

Therefore, thus says the Lord GOD: An adversary shall surround the land, and strip you of your defense; and your strongholds shall be plundered"[39]

The desecration of Mother Earth has deprived us of her nurturing benefits and disconnected us from God and our neighbors. We have turned spirit lines that carry love and divinity into lines that hum with negative consciousness.

Science is Playing God

The thought form of science has not only sucked us further into the physical plane but its fruit is beginning to mutate in Frankenstein-like fashion. Science has started to take on the role of God, or in some cases nature spirits through genetic engineering, bioengineering, cloning and genetically modified food. The problem goes beyond the dependence that a thought form looks to create and begins having science trying to supplant God. These developments are an abrupt acceleration of dependence and threaten a major dislocation.

Early in twentieth century scientist and theosophist Rudolph Steiner understood fully how tampering with nature and her processes could create major disruptions and imbalances. Almost a hundred years ago, before our current honey bee decline, Steiner predicted that scientific and business meddling with bees would be catastrophic:

[W]ith the artificial breeding of bees. You can increase their production of honey, all the work they do, and even the worker bee's capability of accomplishing this work. The only problem is, as Mr. Muller have just stated, that this whole procedure should not be carried out in a way this is too rational and businesses like. Next time we'll investigate more thoroughly the matter of breeding of bees, and we'll see that what proves to extraordinarily favorable measure upon which something is based today may appear to be good, but

145

that a century later all breeding of bees would cease, if only artificially produced bees were used. We want to be able to see how that which is so wonderfully favorable can change in such a way that it can, in time, gradually destroy whatever was positive in the procedure.[40]

As Steiner notes these so called innovations at first seem innocuous but longer term become disastrous. This is the core of the deception created by false gods like science and the market—what initially appears to be such a good idea turns out to be a catastrophe and creates more dependence on them to find a solution.

The editors of Steiner's Bees were even more explicit: "Steiner explained the intricacies of the queen bee, mentioning that the modern method of breeding queens(using larvae of worker bees, a practice that had already been in for about fifteen years) would have long-term detrimental effects—so grave that 'a century later all breeding of bees would cease, if only artificially produced bees were used."[41] Modern breeding methods are disrupting the thought forms, or archetypes, that have guided honey bees ever since they emerged. In trying to make something new science is killing the old.

Gunter Hauk in the introduction to Bees noted that the processes that Steiner had railed against have continued and "that over 60 percent of the American honeybee populations have died during the past ten years."[42] The decline of America's bee population continues and threatens the survival of several crops and our overall food supply. In 2007 the NY Times stated that, "More than a quarter of the country's 2.4 million bee colonies have been lost."[43] As the title of the NY article noted "Scientists Race for Reasons" to determine what is killing the bees, instead they should read Steiner and look in the mirror.

The threats to our food source because of businesses and science meddling with bees pale in comparison to the potential problems created by their pursuit of such things as cloning and genetically modified foods. To better understand this focus, not on the Frankenstein monsters we are creating, but rather on the thought forms behind them, a hideous disregard for nature, Mother Earth and God and the desire to play God. It should be remembered that at its core a thought form of self-interest ultimately looks to prevail and in the process violently harms us, Mother Earth and God's creatures. Science's desire to play God represents acceleration away from God by creating dependence through the destruction of nature and Mother Earth.

The rush to embrace alternative forms of energy that are green to stop global warming such as windmills may be causing unimaginable problems. The blades of large windmills disrupt the flow of the flow of prana creating blockages and even negative vortexes.[44] Since large windmills have been in use for only a short time it is difficult to assess the longer term consequences of the damage that they are doing to Mother Earth. What is clear is that while global warming and pollution can kill us, we cannot live without prana.

It is important to realize that what science promises and looks so appealing in the physical plane initially will always have detrimental consequences longer term. Ultimately, problems develop as we have seen with bees. Scientists scramble coming up with new devices and solutions that ultimately exaggerate the problems longer term thereby creating an even larger vicious circle of dependence and destruction. This truly brings us closer to the Armageddon like calamity many expect.

Separateness

A variety of groups and affiliations vie for our attention. When we affiliate with an organization, club, ethnic group, racial group, political party, nationality, advocacy group, union, corporation and others we are creating or adding to a thought form. The more formalized and structured the thought form the more power that we give it and the more it becomes an artificial construct.

Groups, like all thoughts forms look to perpetuate themselves. Affiliations bring the power of unity, but they also instill the sense of separateness. This separateness ultimately leads to violence either through our thoughts, or through actions. Racism and prejudice are the result of our separateness as is genocide and ethnic cleansing. False identity, or defining ourselves by identifying with some group is the cause for much of the violence in the world today says Christopher Catherwood in *Why Nations Rage, Killing in the Name of God*: "Identity is at the heart of who we are. Yet who we are, and why we are who we are, is often disputed. Identity—nationalist, religious, cultural, and political—is at the heart of much of the conflict in the world today."[45]

The modern world is forming a whole host of new affiliations and groups that separate us. Marketing experts' pigeon hole us through market segmentation by tying our identity to the TV programs we watch, what we eat, etc. in order to sell to us products. Cable TV and other media similarly target us with specific channels and programs that continually give us the same thing. The increasingly heated exchanges in politics and on media are a testament to how we have become fragmented.

Physicist David Bohm understood the danger posed by separate disciplines, separate religions, countries and the like:

Man's natural environment has correspondingly been seen as an aggregate of separately existent parts, to be exploited by different groups of people. Similarly, each individual human being has been fragmented into a large number of separate and conflicting compartments, according to his desire aims, ambitions loyalties, psychological characteristics, etc., to such an extent that it is generally accepted that some degree of neurosis is inevitable, while many individuals going beyond the "normal" limits of fragmentation are classified as paranoid, schizoid, psychotic, etc.[46]

Coming together is a natural process that occurs all the time. The problem arises when we create structures and make permanent (for generations) large groups and affiliations, then groups take on a life of their own.

We should seek unity with God, humanity, nature and Mother Earth. When we unify by race, religion and nationality we are creating separatism and distinction.

Part III Making Heaven on Earth

Chapter 9

Samskaras Can Free Us

If our thoughts have trapped us, then they can also free us. Just as we have created malevolent thought forms and idols we can also create loving thought forms and add to the strength of God. We need to realize that we live in a supernatural world and to begin thinking and acting accordingly. We have it within ourselves to change the world.

Transcendental knowledge is the truth that can set us free. Understanding how to work with thought forms and Mother Earth is empowering and liberating. Such knowledge, if properly applied by a few dedicated souls can be powerfully transformative. Just as technology has empowered a few to move the masses, so can the love of a few with knowledge of Mother Earth and an understanding of the unseen world help heal the many trapped today.

What follows is a guide for transforming the world. We have it within ourselves to bring the Kingdom of Heaven upon earth.

Commitment

The first step in making a better world begins with your commitment to something greater. To see yourself as having a greater purpose and being part of something for the greater good. Many of you have already made a commitment and are well along on your path of spiritual development. You should not abandon your faith tradition, although I would recommend exploring others. All I am asking is that you be on some path.

You should develop your spiritual self, not because you want to advance but because you want to help. Understand that if you neglect your spiritual self you are neglecting God. See yourself as part of the foundation of God's kingdom and as such need to be strong. You will have to carry the weight of many of our brothers and sisters who are too caught up in the material world.

When you dedicate yourself to the greater good and helping others untold benefits will flow to you because what you give will be given back to you many times over in unimaginable ways. My mother always used to say 'give with one hand and get back with two hands.' Remember the law of attraction and other mystical truths.

We need to become bright beacons of God's light and love. In becoming so we bring greater resonance of God to the physical plane and raise the collective consciousness of humanity.

Control Your Thoughts

If you have not already begun, it is important to learn how to control your thoughts and focus your mind in the proper direction. The Haudenosaunee prophet the Peacemaker said that a "clean mind"[1] was one of the requirements to be accepted into the Iroquois democracy, implying an unclean mind was a non-starter for joining and making democracy work. The Buddha tells us that we must control our thoughts, but understood it is difficult, "As a fletcher makes straight his arrow, a wise man makes straight his trembling and unsteady thought, which is difficult to guard, difficult to hold back."[2] Much of Buddhism deals with controlling the mind and spiritual exercises such as meditation. The Noble Eightfold Path of Buddhism is a practical guide on how we should control the mind through wisdom (right view, right intention), ethical conduct (right speech, right action, and right livelihood) and mental discipline (right effort, right mindfulness, and right concentration).

Spiritual practices such as meditation are a necessity. When we meditate we are making a concerted effort to control our mind. It is no different than an athlete who does a physical workout, or weight training to keep in shape to maintain and improve his/her physical body. When we meditate we are exercising our ability to concentrate and focus as well as relieving stress. Most of all we are learning to better control our mind and opening a door of communication with the divine. Meditation also teaches us about our mind and our inner being. By observing our thoughts we begin to see how our mind works. From observation flows knowledge and understanding. Meditation helps us absorb cosmic prana and in doing so raises our consciousness.

The Bhagavad-Gita tells us that, "For him who has conquered the mind, the mind is the best of friends; but for one who has failed to do so, his very mind will be the greatest enemy."[3]

In his translation of the verse Swami Prabhupada says that, "Unless the mind is controlled, the practice of yoga (for show) is simply a waste of time. One who cannot control his mind lives always with the greatest enemy, and thus his life and its mission are spoiled.... As long as one's mind remains an unconquered enemy, one has to serve the dictations of lust, anger, avarice, illusion, etc."[4]

Rama Prasad in his interpretation of *Nature's Finer Forces* says meditation helps burn samskaras and provides a strong counter force to those things that stand as impediments to our spiritual progression by getting us to focus on the divine.[5]

Mindfulness

Meditation is great but we cannot check out and spend all of our lives in a trance. When we are out in the world we are barraged with temptations, pernicious thoughts and false gods. Meditation builds our concentrative ability and helps us deal with and overcome much, but more needs to be done to control our thoughts throughout the day when we are not meditating.

Thich Nhat Hanh recommends practicing Buddhist mindfulness in our daily lives by being continually aware of what we are thinking and doing at all times: "Mindfulness is the practice of stopping and becoming aware of what we are thinking and doing. The more we are mindful of our thoughts, speech, and actions, the more concentration we develop. With concentration, insight into the nature of our suffering and the suffering of others arises. We then know what to do and what not to do in order to live joyfully and in peace with our surrounding."[6]

Mindfulness is like meditation because we are practicing controlling the mind. Being able to be in the present is important if we are to advance in meditation, otherwise our awareness cannot enter the state of timelessness. There are many other teachers that talk about living in the moment and offer a variety of techniques to control your thoughts (See the Recommended Reading List.)

Praying Ceaselessly

Another way to control your mind is to focus on a positive alternative. Pick something close to your heart such as bringing peace and love into the world, or making heaven on earth, or God, and think about it during the day. Treat it as you would a walking meditation.

I would suggest starting by meditating a few minutes on the subject in the morning. For example, ask how you can be an agent for peace and love in the world. Or ask what is love? During the day ponder that question and consider the ways that you can bring about peace and love.

View your experiences during the day as a way to manifest peace and love. Perhaps you decide to give someone a hug instead shaking hands, or you spend time talking to people you usually ignore, or just give everyone a big smile that you meet and send them a heartfelt prayer of love. Look within yourself to determine what you can and should do. Tell your self that you will be a vehicle for peace and love. Explore, contemplate, meditate, focus. Over time you will begin forming a powerful thought form of love and the divine that will begin to evolve in beautiful and unimaginable ways. You will learn subtleties and creative avenues to realize your focus. You will also find love coming back into your life.

Treat your effort as you would meditation. When your thoughts begin to stray bring them back in line by focusing on the subject. No doubt this may be very challenging, as was learning meditation for the first time, but stick with it and keep bringing your focus back. Over time as you develop a powerful samskara it will be easier and easier to stay focused.

You will not be able to keep focused all day. Your job and family life will have to take precedence at times, but you will find that your approach to family, friends and job will become more loving.

You can also repeat a mantra mentally in your mind throughout the day. The story of the Russian pilgrim[7] tells how, in 19th century Russia, a young man sitting in church heard a voice within say to "pray without ceasing."[8] That seed thought began a quest that initially had him exploring his own soul, Christianity and the community around him to determine what it meant to pray ceaselessly. He chose the Jesus prayer, 'Jesus Christ have mercy on me' and would initially repeat it 6,000 times a day. Then it was 12,000 times a day and eventually it was non-stop as he began a pilgrimage to the holy land. Along the way he developed a greater communion with God. Eventually the mantra became self generating and would repeat in his mind with no effort.

We become what we think about and focus on. Repeating a mantra or thinking of love and the divine continually is powerfully transformative. We become love, or take on divine qualities. We also become a bright beacon of God's light and love in the world radiating love, compassion and grace.

You may want to periodically change your mantra or focus. I would recommend being judicious about changing what you are praying on ceaselessly because you want to develop a samskara that will help lift you up. If you are constantly changing your focus it will hamper your ability to create a samskara. Give yourself a minimum of a few weeks, preferably a few months before starting a new prayer.

Shortly I will give you several topics that I ask that you please focus on. Learning to pray ceaselessly can be challenging and initially dangerous if we are not careful how and when we do it. We should be very careful about practicing it in potentially risky situations like driving a car. But ultimately you will find a greater source within you guiding you and transforming your very being.

Meditation, mindfulness, praying ceaselessly and contemplation all increase our ability to tap into and absorb Mother Earth's beneficial essences and in doing so we strengthen our spiritual body.

Nip Negative Thoughts in the Bud

Learn about your mind and thoughts. Become aware how your thoughts affect you physically and emotionally. Notice how your mind works when you are in a hurry and how your breathing rate similarly increases. See how old issues and concerns keep reoccurring in more subtle, or in an even more pronounced manner.

Make an effort not to dwell on negative thoughts and actions since they will have you building negative samskaras that will plant more seeds of negativity. Try to nip negative thoughts in the bud before they can blossom and begin to bear fruit. Observe your negative thoughts and let them pass away—acknowledge them and then let them pass. You might want to label them as you do in meditation, saying that's a negative thought. The key is not to give strength to negative thoughts.

Try to counter any negative thoughts you have with loving or positive thoughts. For example, if you get angry at some one send out love to them as soon as possible afterwards. Patanjali tells us that when the mind is disturbed by passions one should practice pondering over their opposites.[9] Not only does this help reduce the disturbances in our mind but it helps reduce the consequences of the negative thoughts forms you have sent out into the world.

When we are mindful of our thoughts and countering negative thoughts we are not reinforcing the swirl of negative thought forms and false gods circling the world. You are also building a powerful samskara that will better help you control your thoughts.

Choose what you focus on

Whatever we apply thought to we give strength to. So we must be careful about what we are mindful of. For example, while driving you may see a giant billboard with a commercial advertisement. Next to the billboard may be a small tree. See the tree. Try not so much to ignore the billboard as give strength of thought to the tree. As you go about your day look to see Mother Earth, see the positive in others, see God in all. Over time you will begin to see your focus shift.

When we see nature and Mother Earth in the world we give strength to her and increase our ability to sense Mother Earth. Over time this increases our ability to better pick up and absorb her life sustaining aspects (sentience).

Look Within

Our soul is a piece of God within us. When we look within we are looking to God. When we look without, outside at the material world, we are looking at the world of attachment, false gods and malevolent thought forms. When we look without we give strength to them and build a life of addiction, dependence and misery. While all is Brahman, the world is cloaked in maya and full of illusions. When we look at others we must similarly look within to see that God is within them as well.

Lama Yeshe tells us of the futility and problems created by looking without: "As long as we think that a refuge from life's difficulties can be found outside of ourselves, there is no way that we can experience true peace of mind. Certainly the possession of wealth and power is no solution. The high rate of alcoholism, divorce and suicide in the so-called developed nations of the world shows that mere material possession do not satisfy restlessness."[10] The Bhagavad-Gita tells us that one finds happiness within and becomes the perfect mystic: "One whose happiness is within, who is active within, who rejoices within and is illumined within, is actually the perfect mystic. He is liberated in the Supreme, and ultimately he attains the Supreme."[11] Looking within means soul searching and seeking guidance for dealing with life's major and minor problems. I find it helpful to ask my higher self and God, 'Is this of love, is this right, what should I do God?' This centers me, gives me perspective and helps me do the right thing. When we look within we move to a higher standard in guiding our thought and behavior.

Gandhi would spend three days in contemplation and reflection before making a big decision. Similarly Jesus spent 40 days in the wilderness before he began his ministry—his first action was to go into the wilderness for reflection. Native Americans say that we should think about how our actions will resonate seven generations into the future before we act.

Looking within begins a process of revelation and connection to God. We give acknowledgement to something greater than ourselves, we reinforce God, we give strength to soul searching and we begin detaching from the physical plane because we are not giving strength (thought) to it. We also build our spiritual self.

Sri Aurobindo's The Mother tells us that looking within begins our spiritual life:

But to live the spiritual life is to open to another world within oneself. It is to reverse one's consciousness, as it were. The ordinary human consciousness, even in the most developed, even in men of great talent and great realisation, is a movement turned outwards – all the energies are directed outwards, the whole consciousness is spread outwards...

But all who have lived a spiritual life have had the same experience: all of a sudden something in their being has been reversed, so to speak, has been turned suddenly and sometimes completely inwards, and also at the same time upwards, from within upwards.[12]

The Mother goes on to tell how at some point our effort bears fruit and we go from asking to understanding: "One no longer seeks, one sees. One no longer deduces, one knows. One no longer gropes, one walks straight to the goal."[13]

Divine Knowledge

Looking within unveils Divine knowledge and understanding that helps us achieve higher states of consciousness. The Brihadaranyaka Upanishad tell us that all is revealed when we look within and examine the soul.[14]

The Bhagavad-Gita teaches us that meditating on God brings revelation and connection to the divine.[15] The Hindu mystics took this to heart and spent hours on end looking within and reflecting on God. By going into deep states of Samadhi the Hindu mystics where able to attain knowledge about prana, the chakras, the nature of the universe, of Brahman, consciousness, of our purpose and much, much more.

Swami Satyasangananda Saraswati tells us that thousands of years ago adepts gained knowledge that only recently science has begun to realize:

Their search was within; they explored the vast dimensions

that constitute the inner life. Mentally they dissected the body and discovered its subtle essence to be the senses. Through meditation on the senses they discovered the corridors and avenues of the mind. By reflecting on the mind they realized the potential energy that was dormant within. By awakening that energy they discovered consciousness, and by uniting the inherent energy with the individual consciousness they realized that they were indeed intimately connected to and a part of the cosmic consciousness...

Several hundred thousand years later, the unified field theory which physicists talk about uncannily points in the same direction. According to this theory, the entire creation is one composite whole and all life, whether animate or inanimate, manifest or unimanifest, is intimately connected.[16]

Patanjali in the *Yoga Sutras* educates us about the power of concentrative meditation to transcend the physical plane. In Chapter Three, Vibhooti Pada, he talks about siddhis, spiritual gifts or psychic powers that are achieved by concentrating on something. Patanjali tells us that by performing samyama,[17] a deep focused concentrative meditation on an object, higher consciousness dawns.[18] Whatever we perform samyama on reveals itself. For example by doing samyama on our samskaras knowledge of our past lives is revealed.[19] By performing samyama on the form of the body we can make ourselves invisible.[20] Imagine where human evolution would be if we had looked within and not without into the material world and space for the last few millennia?

Looking within has much broader meaning beyond spiritual exercises such meditation. It means seeing God in all and asking what one can do to make a better world. It is about developing a process and mechanism by which we live.

Making Heaven on Earth

Meditation, whether you call it contemplation or focused prayer, is a powerful tool for our individual and collective spiritual development. Hindu's and Buddhists believe that meditation can bring enlightenment, nirvana and end the cycle of death and rebirth. If meditation can bring liberation for one, it will work for our collective consciousness because as we said earlier, as it is for one, so it is for many.

The process of meditation is the process of developing a samskara to help you transcend the physical world. When we meditate we build the impression (thought form) of concentration and of looking within. All the while we are learning to shut out the material world. As we progress in our meditative practice our concentration and focus improve. Overtime we begin to loose contact with the physical world as our physical sensations diminish while we are meditating. Eventually we loose all awareness of the environment and sounds around us while we are meditating. As this process progresses we are all the while burning samskaras of the worldly life.

Swami Vivekananda says that it is only the process of meditation that is capable of shutting out other samskaras, particularly the state of Samadhi: "The samskara raised by this sort of concentration will be so powerful that it will hinder the action of the others and hold them in check."[21]

When we enter Samadhi our consciousness transcends the physical plane and moves to the mental plane, or our manomaya kosha. Swami Satyananda Saraswati in his commentary of *Patanjali's Yoga Sutras* says,

> Samadhi begins only after your consciousness has become free from the physical sphere. The boundary line of the sense world, or maya, ends where pure mental awareness begins. If one is able to withdraw the physical as well as the pranic sense of awareness, but be aware of mental awareness, that is

the beginning of Samadhi. This Samadhi begins when consciousness has gone deep into the manomaya kosha, the mental body, where there is no trace of physical or pranic awareness.[22]

In other words, liberation begins when our consciousness transcends the physical plane and moves to higher planes of existence, but does not reach its ultimate goal until consciousness has transcended all the koshas and resides in the highest plane of existence.

If our own individual liberation begins when our consciousness transcends the physical plane it must be so for our collective consciousness as well; because as it is for one so it is for many, since many are one and one is many. Similarly meditation or the process of detaching from the physical plane, if done with others can bring about our collective liberation.

Samskara can Free Us

If our thoughts can trap us they can also free us. The power of samskaras once established is beyond compare. They move forward with us into future lives as karma. The mystics understood how the power of samskaras could be harnessed for the good. Tibetan Buddhists believe so much in the power of meditation, or our thoughts (samskaras), to transcend the physical plane that they employ various meditative practices to help in the dying process. Sogyal Rinpoche in *The Tibetan Book of Living and Dying* tells how the phowa practice of consciousness transference can facilitate the passage from life to death. He describes a variety of mediations[23] that we should try to make second nature so that we can call upon them at the moment of death. He also notes that the same phowa practices can be employed by others to help the dying person, even if the dying person is not practicing phowa themselves.

The same principal of creating samskaras with the phowa

practice can be applied constructively to help us collectively transcend the physical plane. Through meditative practices and concentrated action we can create a powerful samskara(s) that can help heal and transcend the physical plane. Further the creation of such a samskara does not require the participation of everyone to get it going.

While the false gods and malicious thought forms have been bearing fruit and casting new seeds of dependence for generations they can over time be shut out. It is possible to construct a thought form that would provide a positive alternative. A samskara focused on love and God that could begin to raise our collective consciousness. Swami Satyananda Saraswati in his commentary of *Patanjali's Yoga Sutras*(I.50) tells us: "Even in ordinary life it is found that when one thought predominates in the mind, the others are subdued. Likewise, one samskara can prevent other samskaras."[24]

So it is possible that if we could create a strong enough thought form or samskara of love and God it could overtake all the horror manipulating and pulling at us and set us free.

While such a task is challenging, it is not insurmountable. Knowledge of how the mystical world works will go a long way to overcoming challenges. The power of prayer and the creation of thought forms can be harnessed to facilitate the process. Knowledge of how Mother Earth works can be strategically applied to spiritual practices to have exponential affects. Hindu mystics have demonstrated how focused concentration on various chakras or on various concepts brings knowledge and develops certain attributes, particularly if done by someone that is spiritually developed and has strong concentrative abilities. Similarly choosing the location of our meditations, or how we do them can greatly enhance them.

The power and force of unity can be applied to the greater good. Dedicated groups of people working together, whether they are in close proximity or not, can have a powerful healing affect on our collective psyche. As this process develops others will be swept up by the whirlwind of light and love of this thought form.

Heavenly Meditations

Since the beginning of the twentieth century a lot of people have learned techniques of visualization and focused thinking to put their thoughts to work. Earlier we had talked about how prosperity preachers and new-agers were employing such techniques to bring abundance and wealth into their lives. Those same techniques can be applied to the greater good. Instead of thinking and contemplating on things that could benefit you, meditate on those things that heal and make the world better.

We need to create a samskara (thought form) that benefits all of humanity and the world. Many of us already do this when we pray for peace, justice and the like.

It would be ideal to have one positive thought for people to focus on, but given the state of the world this is not an option. I am proposing four basic thought forms that people can focus on to bring heaven upon earth. They are all related. They are as follows:

God: since the beginning of time we have prayed to, worshipped and looked to God. Almost universally the prophets and sages have said we should pray and look to God, whether you call it Spirit, Source, Creator or the countless other names. Because of this, there already is a powerful samskara for God in the world that we can add to.

Love: love is at the core of who we are and how we are to live. It is the antithesis of the market's dogma of self-interest. Jesus told us all flows from love.

Mother Earth: because we have desecrated Mother Earth our very existence is threatened. Wounded and covered with our negativity and violence, she is unable to pulsate all the love, sustenance and divine attributes we need, thereby making us vulnerable. There is much to be gained with meditation and intention on Mother Earth, that it's almost inconceivable all the good that can be derived. By praying for her and reclaiming sacred space, we can revive and build a bond that can help us transcend the physical plane.

Heaven on Earth: it is a possibility and we need to realize it and believe in it. The misguided and negatively focused preach of Armageddon and tell us all is lost, but they are wrong. Heaven on earth is a real possibility and part of God's divine plan. What would such a beautiful world look like? Just imagine and contemplate a beautiful world where peace reins, happiness and joy permeates the air and old enemies lay down next to each other like the lion and lamb.

Remember that whatever we give thought to we give strength to and whatever we think about we become. By focusing on these four meditations we become powerful ambassadors for them.

Be the change we wish to see

It is not enough that we mediate and pray on these four medita-tions. We must take them to heart and make them part of our life in all that we do. Our thoughts, actions and speech need to be reflective of and give strength to these meditations. As with praying ceaselessly we must make thought, speech and action reflective of the prayer. We must be the change we want to be.

Many associate Mahatma Gandhi with the statement that "We must be the change we wish to see." Scholars who have written biographies of Gandhi[25] have told me that they were not able to find where Gandhi said this. In his *Satyagraha* Gandhi talks about merging the mind with action.[26] I believe that this phrase may have been coined by Sri Aurobindo's the Mother, Mirra Alfassa, in speaking about the spiritual life and how it can become a contagion that can sweep others up with its radiating love: "If one sincerely wants to help others and the world, the best thing one can do is to be oneself what one wants others to be – not only as an example, but because one becomes a centre of radiating power which, by the very fact that it exists, compels the rest of the world to transform itself." [27]

Both Sri Aurobindo and the Mother were contemporaries of Gandhi and spent a lot of time in spiritual efforts to win freedom for India, end to WWII and bring about world peace. Sri Aurobindo was a nationally known political activist in his youth who headed the underground Jugantar party that advocated independence for India before his spiritual rebirth.

Similarly Thich Nhat Hanh says that when we marry mind, body and speech much is possible: "We have to pray with our body, speech and mind and with our daily life. With mindfulness, our body, speech, and mind can become one. In the state of oneness of body, speech, and mind, we can produce the energy of faith and love necessary to change a difficult situation." [28] When we unite mind, action and speech on a particular thought or effort we are focusing our attention as we if we are meditating. It is such focused attention that gives power to the effort. Just as with meditation, when we become absorbed with what we are thinking about distractions melt away, knowledge and understanding flow. The concentrative effort of uniting thought, action and speech on one particular effort or thought creates a very powerful thought form. When someone who has developed their spiritual strength focuses their thoughts, actions and speech on one thing, much is possible. When several people do it anything is possible.

Merge action with thought

It is merging of action with thought (intention) that makes a ritual or ceremony powerful. Take for example, the Christian communion. Jesus tells us: "While they were eating, he took a loaf of bread, and after blessing it he broke it, gave it to them, and said, 'Take; this is my body.' Then he took a cup, and after giving thanks he gave it to them, and all of them drank from it.

He said to them, 'This is my blood of the covenant, which is poured out for many."[29]

It is the process of symbolically eating bread that is the body of Christ and drinking the wine that is the blood of Christ that give power to the ritual. These physical actions reinforce the taking of Jesus into your life. Similarly the Native American purification ritual or sweat lodge relies on the physical action of sweating to reinforce the process of purification. You are trying to purify yourself mentally, spiritually and physically.

We must practice what we preach and try to make sure that our words, actions and thoughts are one.

The Mystical Body of Our Collective Conscious

The idea of a collective conscious for good is a concept held by many faiths. Buddhists call an assembly or community of like minded people with a common goal and purpose a sangha. Often sangha refers to a monastic order. Thich Nhat Hanh tells us that the collective consciousness of a sangha has a powerfully transformative effect:

Right now, the collective consciousness of society is in very bad health. But we can learn to heal and transform ourselves. To do this we have to create a Sangha body, that is, a collective consciousness that is able to protect us. In the cities, you only need to look at the sights, listen to the sounds, and be in touch with a small number of people, and you can fall sick. When you come to a retreat space, you can close the door on all of that and open the door to the spiritual realm. Your body receives a physical, as well as a mental, boost.[30]

While he is referring to a physical sangha, such as a monastery or retreat center which certainly can be a powerfully transformative place, the idea of a collective consciousness applied to a common purpose or vision not physically domiciled in the same place, can similarly be transformative. As was discussed in Chapter 2, "Thoughts, Thought Forms, Samskaras" and demonstrated throughout, our thoughts do not necessarily need physical proximity to be powerful. What creates a contagion of thoughts is its like mindedness. As Thich Nhat Hanh notes the sangha has the ability to purify and create consciousness and energy of such magnitude that it can even change our karma (samskaras.)[31]

The spiritual strength of the community members will give strength to our collective consciousness and the focus on the four meditations.

Swept up in a Whirlwind of Love

It is not necessary to have everyone focus on the four meditations. No doubt many will be dismissive and unbelieving. However, they can be swept up by the same sort of whirlwind of thoughts for love and God that swept them up into worshipping the market and science. Many are ardently devoted to the market and science and subsequently have not developed their spiritual strength. Not having spiritual strength leaves one open to all sorts of things and this why so many to fall as prey to malicious thought forms and false idols. This also makes them open to the positive—love, compassion and God.

Others are misguided, believing that the world is fair and balanced, or believing in a God of wrath. Education and understanding will help them see.

As people take the four meditations to heart it will create a very positive and powerful beacon of light that will shine upon other souls and bring them closer together.

What follows is a detailed explanation of the four mediations and how you can implement them in your life. How you choose to do this is up to you. Your background, faith and spiritual development will determine how you incorporate the four meditations into your life.

Chapter 10

God

There are many ways that we can recognize and give strength to God. It does not matter how you perceive of God or whether you call God, God, The Creator, Source, Spirit, Allah or countless other names. The important thing is the realization that there is something greater in the universe no matter what you call it, or how you perceive of it. As we noted earlier each thought has a generic nature or character to it so it does not matter whether we focus on God or Creator.

All faith traditions prescribe various spiritual exercises such as prayer and mediations as well as right ways of living and loving our neighbor to be in harmony with and reaffirm God. Those are important concepts and practices that need to be addressed. First we need to ask, if God has been worshipped since the dawn of civilization why are the market, science and other malicious false gods and thought forms so dominating our world?

One would assume that all the focus on and the worship of God over the ages that it would give such strength to God that God would dominate, yet the market, science, technology and other false gods and samskaras rule. Part of this has to do with our being trapped in the physical plane of senses and pleasures and being enraptured with them. The primary reason is that we have failed to believe in and follow a God of Love that wants compassion and mercy from us and for each other. Instead of seeing a God of Love much of recorded history has been dominated by the belief in a violent and vengeful God of Wrath. The thought and action applied to such a God of Wrath has given strength to violence, vengeance and much of what is wrong with the world today. Since time immemorial we have killed in the name of God and usurped God for our own ends.

We need to believe in a God of love and incorporate a God of love into our lives through our thoughts, actions and speech. When we do this we reaffirm God and give strength to God, and in so doing lessen the divide between us and God.

A God of Love

A God of love is full of love, gives love and wants us to love each other. As we noted earlier the golden rule of do unto others is universal to all religions. Jesus told us that all flows from the law of love.[1] In other words it is the core teaching given to us by God. Mohammed tells us that Allah is, "full of mercy and loving kindness."[2] 1 John tells us, "Whoever does not love does not know God, for God is love... God is love, and those who abide in love abide in God, and God abides in them."[3] Isaiah tells us that God has unswerving love for us: "For the mountains may depart and the hills be removed, but my steadfast love shall not depart from you, and my covenant of peace shall not be removed, says the LORD, who has compassion on you."[4]

To many this proclaimed love seems at odds with a God that would blot out much of humanity as was done at the time of Noah. Why did God bring the flood and not "turn the other cheek"[5] as Jesus advocated?

To see a God of love we need to understand that the rule of karma operates in the physical plane and that if you kill you will be killed. For example in Exodus God specifies how what you do will determine what will happen to you: "You shall not wrong or oppress a resident alien, for you were aliens in the land of Egypt. You shall not abuse any widow or orphan. If you do abuse them, when they cry out to me, I will surely heed their cry; my wrath will burn, and I will kill you with the sword, and your wives shall become widows and your children orphans."[6]

Yes, God can give us grace and others can take our karma on, but ultimately karma prevails.

God has sent us prophets and great teachers and is constantly imploring us to walk the path of love.

God of Wrath

Unfortunately the greater part of humanity has neither believed in nor followed the path set by God. God through Hosea tells us that," For I desired mercy, and not sacrifice; and the knowledge of God more than burnt offerings."[7] Similarly Jesus tells us, "I desire mercy, not sacrifice."[8]

But history is not filled with justice, compassion and mercy. Quite the contrary it is filled with violence, hate, racism, sexism, slavery, violence and war often done in the name of God. We have had crusades, jihad, holy wars and more killing in defiance of God's call to love each other, much of it sanctioned by religious authorities.

The self-righteous and violent see God as being wrathful, a God who doles out violence to the sinner and evil doer. They see the world as the battle for good and evil. They are the good, the enemy is the evil. They justify their violence by believing that they are doing God's work by ridding the world of evil. They often define their faith by condemning (judgment) and even killing their enemies. They kill in the name of God because they fail to see the God in all, that my enemy is my brother, my sister. Their simplistic vision ignores God's plan for everyone's spiritual transformation on the physical plane. They do not see that our individual spiritual evolution lies in a continuum between the divine and the demonic and we are here to be transformed to the divine. Christian fundamentalists often take pride in Paul the apostle and his teachings in the New Testament, yet they forget that before his transformation he was the zealot Saul of Tarsus who oversaw the persecution of many Christians.

Redemptive Violence

Devotees of a God of Wrath see redemption through violence, that by killing the bad guys they will get to heaven. As Walter Wink notes the idea of redemptive violence permeates our lives:

> The myth of redemptive violence inundates us on every side. We are awash in it yet seldom perceive of it… [I]ts simplest, most pervasive, and most influential form, where it captures the imagination of each new generation: Children's comics and cartoon shows.
>
> Here is how the myth of redemptive violence structures the standard comic strip or television cartoon sequence. An indestructible good guy is unalterably opposed to an irreformable and equally indestructible bad guy. Nothing can kill the good guy, though for three-quarters of the strip show he(rarely she) suffers grievously, appearing hopelessly trapped, until somehow the hero breaks free, vanquishes the villain, and restores order until the next installment.[9]

Redemptive violence implies that killing is justified because we are not only doing God's work but we will be rewarded for it. Advocates of redemptive violence erroneously believe that killing is good. In such a fantasy state anything and everything is justified, law breaking, killing, lying and the ten commandments are a put on hold because the ends justify the means. This is in sharp contrast to what God teaches. God wants us to do to the right thing and love our neighbor and our enemy. Resorting to violence as a means to achieve a certain result is wrong, even if you believe it is for the greater good. We should, as the Bhagavad-Gita teaches, not be concerned with the fruits of our actions.[10] Do the right thing first and foremost.

Killing is killing. Murder is murder. No matter whether you cloak it in God, the greater good or self defense, murder is murder. Killing in the name of God creates a very powerful and negative thought form. Killing is the ultimate form of disrespect towards God because it reflects the antithesis of love and compassion and brings the finality of death. Both the self-righteousness and emotion of killing gives added strength to the thought form of violence and murder.

Scripture has been twisted by the devotees of a God of Wrath to justify violence as seen with Augustine's Just War Theory, the Crusades, or recently with President Bush's invasion of Iraq and Osama Bin Laden's 9—11 terrorist acts. Both Bush and Bin Laden hijacked God to justify violence. When leaders and theologians extol violence and demonic behavior they reinforce the power of violence and undermine the belief in a God of Love and reinforce the belief in a God of Wrath.

Wink believes that the faith in redemptive violence has created our 'Domination System' of perpetual war, where peace is the achievement of war and prosperity the fruit of a successful war.

Le Bon tells us that when a crowd forms it has singular purpose and participants are willing to sacrifice for a greater purpose. In a world of violence ruled by a God of Wrath, the hero is the solider. The solider kills and risks his life for a greater purpose in the world of violence. It is sharp contrast to the example of Jesus who had Peter lay down his sword when they came to arrest him and voluntarily let him be taken away to his death on the cross. It is sad to think that so many have died thinking that their sacrifice and violence was making the world better. Unfortunately their good intentions and willingness to sacrifice is overwhelmed by the darkness of the seed thought of violence and murder that they are festering.

We cannot love if we see God (others) as being evil.

Our Violent World

It is the violence and misery in the world that has so many believe that God is wrathful and vengeful. We ask ourselves how

God could let such horrific things such as genocide, torture, murder and disease happen?

We need to realize that it is not God that is doing these things but mankind. People are killing people. Mother earth is recoiling with storms from the abuse we have done to her and each other. It is our violent and negative thoughts and actions that have made the world what it is. We are living in the soup that we have made.

Aldous Huxley in the *Perennial Philosophy* says that it is our separation from God and our misperceptions about our unity that makes for a violent world that can even ensnare the righteous: "[M]ost human are chronically in an improper relation to God, Nature and some at least of their fellows. The results of these wrong relationships are manifest on the social level as wars, revolutions, exploitation and disorder; on the natural level, as waste and exhaustion of irreplaceable resources."[11]

The challenge is to understand that those committing the violence and hate are like us, part of God. Scripture speaks to this metaphorically when it tells us how challenged we are to see God. Moses could not look into the burning bush to see God. Similarly Arjuna was shaken to his core when Krishna revealed him(her)self: "O mighty-armed one, all the planets with their demigods are disturbed at seeing Your many faces, eyes, arms, bellies and legs and Your terrible teeth, and as they are disturbed, so am I. O all-pervading Vishnu, I can no longer maintain my equilibrium. Seeing your radiant colors fill the skies and beholding your eyes and mouths, I am afraid."[12]

Arjuna would later ask to see the conventional form of God he was comfortable with, "O universal Lord, I wish to see You in Your four-armed form, with helmeted head and with club, wheel, and conch and lotus flower in Your hands. I long to see You in that form."[13] To see and live God we have to love all and accept the many faces, arms and bellies. When we see God in all we reaffirm and give strength to God.

Getting closer to and Connecting with God

Mohammed tells us that we are close to God, "We are nearer to him than his life(jugular) vein".[14] It sure does not seem so. Many of us may be close to God, but most of us are pretty disconnected.

Getting closer to God is about seeing God in all, to feel divinely inspired, to look within and to hopefully hear God speak to you in different ways. These are a few of the qualities that you are looking to develop. As with meditation and all other spiritual exercises, getting closer to God means developing a samskara that will help you bridge that connection and develop a strong bond. When you get closer to God you give strength to God's presence in the physical plane and increase your ability to connect with God. As you nurture your communication with God you are growing that thought form of divine connection that will enhance the ability of other's to also connect with God. Developing the divine connection is about shifting the focus from the material plane to higher planes of consciousness.

There are many ways that you can begin this process by. Spend some quiet time reflecting on God and your relationship with God. Ask God to speak to you. Make this a regular part of your life.

Look within for solutions in your life. Don't simply react or do something because it is the law. Ask God what course you should take.

Give thanks. Native Americans have a wonderful tradition of giving thanks that all of us should incorporate into our lives. Give thanks for the big and little things that you feel blessed about. Although it can be difficult to say thanks for the bad things, or challenges you face. Say thanks for the life lessons that adversity brings; thanks for how it could have been much worse, or thanks for given the opportunity to learn. This is not a call to be masochist. Some tragedies are too overwhelming. It is meant to help you develop a positive attitude and perspective with your life through challenges and triumphs. When you begin to see the silver lining and blessings in adversity you become indifferent to what comes your way and just feel blessed to be. In many ways you will be walking what Buddhist's call the middle path or the Hindu's teach about being indifferent to the fruits of action.

Giving thanks can be spiritually as well physically transformative. You will find the added benefit of joy and happiness comes into your life from giving thanks because you will become appreciative of life.

Sincerity

It is your earnestness and sincerity in wanting to connect with God that can facilitate the process. Jesus tells us to be solemn, not to be public and to be genuine when we pray.[15] If our intentions are pure and we apply ourselves then at some point God will speak to us. The history of manifestations of divine beings on the physical plane is often associated with the innocent and the devout, Marie-Bernarde Soubirous (St. Bernedette) at Lourdes, the three children at Fatima and others that have been blessed with seeing apparitions because of the sincerity of their devotion.

Earnestness can be achieved through effort. If we passionately apply ourselves to communicating with God we will eventually be rewarded. You can try a mantra of pleading or requesting that God speak to you.

In his biography *Black Elk Speaks* written by John Neihardt, Black Elk describes the lengths that the Sioux would go to get a vision, cold nights on a lonely mountaintop, lamenting and crying for God. In the *Sacred Pipe* Black Elk tells us the importance of lamenting:

The 'Crying for a Vision'...this way of praying is very important, and indeed it is at the center of our religion, for from it we have received many good things...Every man can cry for a vision, even 'lament'; and in the old days we all —men and women—'lamented' all the time...

[T]he most important reason for 'lamentation' is that it helps us to realize our oneness with all things, to know that all things are our relatives; and the in behalf of all things we pray to Wakan-Tanka that He may give to us knowledge of Him who is the source of all things, yet greater than all things.[16]

Similarly Sri Ramakrishna would spend countless hours or even days praying ceaselessly to the Goddess Kali. As biographer Christopher Isherwood noted, "Ramakrishna, as an exemplar of spiritual practice, had to demonstrate, for all of us, that devotion can also lead to unitive knowledge."[17] It is said that Ramakrishna with the wink of an eye could go into Samadhi and stay there for hours, days and even weeks on end.

Seeing God in all

When we begin to see God in all things we move closer to God and having heaven upon earth. Learn to look at the world around you and see God in all things. Jesus tells us when we feed the poor, welcome the stranger and visit the prisoner we are helping God: "[F]or I was hungry and you gave me food, I was thirsty and you gave me something to drink, I was a stranger and you welcomed me, I was naked and you gave me clothing, I was sick and you took care of me, I was in prison and you visited me...Truly I tell you, just as you did it to one of the least of these who are members of my family— you did it to me."[18]

As much as you are helping others you are helping God and yourself.

Realize that God is always speaking to you and reaching out to you. Listen and look for it. God is with your neighbor, your enemy, the stranger and the ones closest to you. The voice of love and compassion, the voice of reason is always close at hand.

Believe in it.

Understand that the divine is always there ready to help and guide you. Believe in it and trust it. As you do your path will become clearer.

Forgiveness becomes easier when we see our neighbor and our enemy as being part of God. Look at them and see God. Don't see their hate, violence or greed being who they are but rather their vulnerabilities, their fears, their insecurities. Pray for them. Even if you find it difficult to believe in what you are saying, still pray, longer term you will come to believe in what you are praying for.[19] Ultimately your prayer will bring understanding to you and help melt a little of that what ails them.

As you see God in all and with everything, God will come to you. Remember the law of attraction. If you put thoughts of God in everything, then everything that comes back to you will be of God. The world will begin to radiate God in your life and the life of others.

Develop your faith

Faith in God is like a muscle that needs to be developed. Believe in it.

Begin slowly with small steps and build upon it. Again you are looking to develop a samskara, this one is your faith. Small steps are the key, too many people incorrectly think that their faith can move mountains immediately.

Trust in God. Be an optimist and see the good in everything. Always look for the silver lining no matter what. Feel blessed by what has happened and give thanks to God.

Do not let fear take hold. Rely on God to get you through challenging times. Take strength from God to help you overcome. When fear comes remember the words of Psalm 23: "Even though I walk through the valley of death I fear no evil; for you are with me; your rod and your staff— they comfort me."[20] Make an effort to regularly visualize God and the prophets being with you and guiding your path in both good and bad times.

You might want to develop a certain mantra that you can use for challenging times. Such as the Jesus prayer, the Lord's Prayer or a Hindu mantra. It is best if the words and intention are about God. Work and develop this prayer. Do it at good times so that you will associate with good times, that way it will better assist you at time to not be overcome with fear. As the Bhagavad-Gita instructs us: "Being freed from attachment, fear and anger, being fully absorbed in Me and taking refuge in Me, many, many persons in the past became purified by knowledge of Me—and thus they all attained transcendental love for Me."[21]

Give Strength to God

Through our prayers and focus on God we give strength to God. The Bhagavad-Gita tells us to meditate on God; "one should meditate upon Me within the heart and make Me the ultimate goal of life."[22]

Take time in your day to pray to God. There are a variety of ways to do this. You can give praise to God, through words, in song or readings of scriptures. Begin activities both privately and publicly with a prayer. If only for a moment, give pause for reflection. Community gatherings whether they be formal services or an informal satsang are a great way to recognize God.

Devotion

You should ultimately strive to devote your life to God. To dedicate your life to serving God and facilitating the kingdom of heaven upon earth. To as Islam means submitting your will to God's will, or as the Lord's prayers says "thy will be done".

The path of karma yoga[23] followed by Gandhi and countless others is a path of living for God. Followers of this path renounce the materialistic world and dedicate all of their actions to God. In other words they are devoted to God and the fruits of their actions are done for God. By detaching from the physical plane and having no attachment to their actions or objects, they are freed.[24]

The Bhagavad-Gita tells says that until we submit our will to God and dedicate ourselves to God we will not be freed: "The steadily devoted soul attains unadulterated peace because he offers the result of all activities to Me; whereas a person who is not in union with the Divine, who is greedy for the fruits of his labor, becomes entangled."[25]

Bearing Witness

We reaffirm God through the action of bearing witness and helping others. You bear witness by pointing out to someone, a group, an organization or government that they are sinning or defiling God through their actions or words. The key is to reaffirm the positive, God or scripture, when pointing out the transgression.[26] You are making an effort to reaffirm the positive and educate and not be against something. Jesus teaches us that we are to point out the transgressions of others against us.[27]

Bearing witness is about being true to yourself and making God wholly part of you. It is about putting God's message into practice and becoming God's servant. Quakers have a long history of standing up to injustice and bearing witness as Michael Birkel notes: "Quaker spirituality is both inward and outward. Friends have always expected the Holy Spirit to transform individuals and then guide them into ways of trans-forming society. The mystical stream in Quakerism has a profound ethical dimension. In worship together Friends have expressed not only wordless union with God but also practical leadings to engage in concrete actions."[28]

Bearing witness is about reaffirming God in all that you do. You have to be more than a navel contemplator, do more than attend religious service if you want to be true to God; otherwise you are part of the problem and not the solution. God's words have to be put into action. This is not meant to say that you have to put your life on the line, rather that you should attempt to incorporate God into all aspects of your life. Action can speak louder than words. Bearing witness merges action with intention. It reinforces God. Look within yourself to see what you feel called to do and build upon it.

Quaker Gray Cox says that,

The guiding concern of people bearing witness is to live rightly, in ways that are exemplar. Insofar as the end they aim at, it is perhaps most helpful to think of it as the aim of cultivating their souls and converting others. They are not so much trying to find a way to get peace as bear witness to the conviction that there is no way to peace, peace is the way. And this way of peace is one of bearing witness to those truths found in clearness, when impulses are quieted and leadings are followed in the gathering of consensus.[29]

Make peace the way by bearing witness.

When we bear witness we publicly point out the transgressions of institutions and educate the public. Barbara Rossing feels that corrupt institutions need to be put on trial.[30] Walter Wink says that many of us are blinded to the reality of the world around us because the powers have created a 'delusional system' that is like a gamed played upon us. Because of this Wink feels that it is our responsibility to 'unmask the domination system' and show the reality of who the powers are; "The struggle for a precise "naming" of the Powers that assail us is itself an essential part of social struggle."[31] According to Wink bearing witness will help the scales fall from the eyes of our brothers and sisters. Bearing witness is very similar to the process of meditation where we watch our thoughts and label them for what they are, only here it is in the context of the larger collective of our group consciousness.

Non-resistance

Sometimes speaking truth to power means that we have to do more than speak out and carry around a placard. In those instances governments and power structures need to be confronted directly with what many call non-resistance, or not partaking in what is evil. Leo Tolstoy says that 'nonresistance' comes from 'resistance not evil, Matt v.39. and is a 'high Christian virtue'. According to Tolstoy "Evil is to be resisted by all just means, but never with evil."[32] As examples, Tolstoy says we should not pay taxes to a government that engages in war and avoid conscription no matter what the cost.

Mahatma Gandhi felt that we can bring about change in the system through civil disobedience:

[O]n the political field the struggle on behalf of the people mostly consists in opposing error in the shape of unjust laws. When you have failed to bring the error home to the lawgiver by way of petitions and the like, the only remedy open to you, if you do not wish submit to the error, is to compel him by physical force to yield to you or by suffering in your own person by inviting the penalty for the breach of the law. Hence Satyagraha largely appears to the public as Civil disobedience or Civil resistance. It is civil in the sense it is not criminal.

The lawbreaker breaks the law surreptitiously and tries to avoid the penalty, not so the civil resister. He ever obeys the laws of the State to which he belongs, not our fear of the sanctions but because he considers them to be good for the welfare of society. But there come occasions, generally rare when he considers certain laws to be so unjust as to render obedience to them a dishonor. He then openly and civilly breaks them and quietly suffers the penalty for the breach. And in order to register his protest against the action of the law givers, it is open to him to withdraw his co-operation from the State by disobeying such other laws whose breach does not involve moral turpitude.[33]

All actions of non-resistance should be centered on God and be accompanied with prayer, contemplation or meditation.

Look within and make God part of all that you do and think.

Chapter 11

Love and Community

Love is living and reaffirming God's message to turn the other cheek, to love your enemy and seeing God in everything and everyone. God is love.

All flows from love. From love comes compassion and mercy.

Love is about creating unity, bringing together and not being restricted by artificial walls and constraints. We need to see the unity in all—between us, with God's creatures and mother earth. Love is the means, the process, by which we unite.

Lao Tzu teaches us that our life should be dedicated to others. His words invoke the words of others, the suffering servant[1] he who is last is first[2] and tells us that loving should not be conditioned by our neighbor's behavior, "Live for others. The universe is everlasting. The reason the universe is everlasting is that it does not live for Self. Therefore it can endure. Therefore the Sage puts himself last and finds himself in the foremost place; Regards his body as accidental and his body is thereby preserved. Is it not because he does not live for Self. That his Self is realized."[3]

Love is about living for others, it is sacrifice, and it is about making your life's purpose something greater than oneself. Krishnamurti says that "Where the self is, love is not."[4] As long as we focus on the 'I' and the 'me', love will be illusive.

Love is the positive alternative in a world caught up in the negative swirl of 'self interest' and its malicious fruit of greed and violence. We need to create a contagion of love that can sweep everyone up in its bliss.

All Flows from the Law of Love

Love is the guiding principle by which we should live. We should endeavor to think and act out of love. We can facilitate this by asking ourselves whenever we do something whether we are doing so out of love and is it in keeping with the law of love. Reflection gives power to the thought forms of love and justice and strengthens our own nature by the intention of wanting to do good. Better yet meditate and think about how you can put forth God's love in the universe. Over time this will become second nature.

Love is the foundation of all scripture. Jesus tells us that all flows from the law of love.[5] The Prophet Mohammed similarly teaches that all the laws and words of the prophets hang together. Implicit in Islam is the concept that each prophet must agree with the previous prophets, lest they be a false prophet. By agreeing with what has been said before they corroborate it and strengthen it. Similarly those quoting or sanctifying their acts with scripture must insure that their intent is in keeping with what has been said before. In other words, does it flow from love?

A thought seed not planted in love or God starts a downward spiral. When Jesus tells us to turn the other cheek he is telling us not to reinforce violence by responding to violence with violence. If we respond to violence with violence, no matter how much we feel justified in doing so, we are reinforcing violence. While you may feel justified in striking back you need to realize that the law of karma is at work. The violence you are experiencing could be a payback from a previous life, or it could be your pre-reincarnation decision to take on the violence of the world as your gift to the world in this lifetime so that you may grow. It all gets rather complicated, but ultimately the law of karma rules and everything balances.

Love and turn the other cheek also implies nonviolence, or ahimsa. Jesus personified non-violence through out his life. When they came to arrest him and Peter fought back and cut off a soldier's ear Jesus told him to lay down the sword and healed the solider. Early church leaders such as Origen echoed Jesus advocacy of peace and nonviolence and refused to take up the sword.[6] Early Christians were so passionate about advocating love and peace that they often accepted their death instead hurting someone else.[7] Maximilian of Tebessa was beheaded in 295 because of his refusal to be conscripted into the Roman army.

Buddhism similarly advocates nonviolence; "one should not kill nor cause another to kill."[8] Jainism not only advocates nonviolence towards others but to all living creatures:

Control of speech, control of thought, observing the ground in front while walking, care in taking and placing things or objects, and examine the food in the sunlight before eating/drinking are five observances of nonviolence... Benevolence towards all living beings, joy at the sight of the virtuous, compassion and sympathy for the afflicted, and tolerance towards the insolent and ill-behaved are the right sentiments."[9]

Ahimsa, nonviolence, can be a powerfully transformative process especially when confronting violence and murder. Gandhi and King showed us this. By resisting violence with love we really do God's work by not reinforcing violence and facilitating transformation.

Justice, Mercy and Compassion
From love flows justice, compassion and mercy. If we are one (unity) and are called to love one another, then we have to strive for justice in the world and have compassion and mercy for all, especially those injured. Isaiah tells us, "learn to do good; seek justice, rescue the oppressed, defend the orphan, plead for the widow."[10]

Rabbi Heschel tells us that, "Righteousness goes beyond

justice. Justice is strict and exact, giving each person his due. Righteousness implies benevolence, kindness, and generosity...Justice dies when dehumanized, no matter how exactly it may be exercised. Justice dies when deified, for beyond justice is God's compassion. The logic of justice may seem impersonal, yet the concern for justice is an act of love."[11]

Similarly the Bahaullah tells us how justice and mercy resonate with God: "O thou who treadest the path of justice and beholdest the countenance of mercy! Thine epistle was received, thy question was noted, and the sweet accents of thy soul were heard from the inmost chambers of thy heart."[12]

When we give love, seek justice and have compassion for others we are doing so to God, because we are one. We are also giving strength to and defining God as being a God of Love.

Be An Ambassador of Love

Become an ambassador of love in your life. Give love to everyone and everything. Take it upon yourself to bring joy and love into people's lives by looking for ways to bring happiness and joy to those that you know, or don't know. Explore what you can do to help others.

As you go about your day look to spread happiness to others. Be friendly and upbeat with everyone that you meet. Ask God to help you in bringing forth joy.

When you find yourself a stranger in a crowded place, like in a bus or subway, send out love. Select someone in the crowd that you do not know, that looks like they could use love, and pray for them. Do not stare at them or acknowledge them in any other way. Just pray with your earnest emotion and love.

Meditate on love. Reflect on how you can spread love, bring love and give love. Be like St. Francis of Assisi who asked God to make him an instrument of peace, and ask God how you can become a bright beacon of God's love and light upon the earth. Visualize yourself self as such a bright beacon. See yourself radiating love and light to everyone and everything.

Buddhist's Christ like compassion

Christians believe that Christ died to bear our sins: "He himself bore our sins in his body on the cross, so that, free from sins, we might live for righteousness; by his wounds you have been healed."[13] Many Christians take that as an example to bear hardship and violence and turn the other cheek; "For to this you have been called, because Christ also suffered for you, leaving you an example, so that you should follow in his steps."[14]

Tibetan Buddhists practice Tonglen meditations by visualizing taking on others suffering and giving others their happiness and joy. It is said that Geshe Chekhawa in the eleventh century popularized Tonglen when previously incurable lepers practicing Tonglen were miraculously cured by the practice.[15]

Sogyal Rinpoche describes a variety of Tonglen practices in *The Tibetan Book of Living and Dying* to increase our compassion: remembering a fond love of our youth and sending it to others, considering ourselves as others, exchanging ourselves for others, using a friend for compassion, meditating on compassion and directing your compassion.[16] Other practices to develop compassion would be to imagine ourselves as a loving deity, or while in meditation visualizing various acts of compassion such as hugging someone you don't like, or turning the other cheek when struck, or sending love to someone that has just done something bad.

Because we become what we think about strong meditative practices of Tonglen will most certainly build compassion within us and possibly help us heal as the lepers under Geshe Chekhawa did. How much compassion practices will help eradicate the karma of those that we wish to help will depend upon our spiritual acumen, the person we are sending compassion to and other factors. Meditative techniques focusing on compassion such as Tonglen can, if done by enough people, have a powerful affect.

Community over charity

All faiths call us to give to and help the poor. The third pillar of
Islam is zakat, or alms-giving. Sikh religious services always
conclude with a community meal and the feeding of the poor.
The Rig Veda says, "On the high ridge of heaven he stands
exalted, yea, to the Gods he goes, the liberal giver."[17] It seems
that almost every page in the Old Testament talks about helping
the poor.

Christians are instructed to give in anonymity and not make
the act a badge we pin on our chest; "[W]hen you give alms, do
not let your left hand know what your right hand is doing, so that
your alms may be done in secret; and your Father who sees in
secret will reward you."[18]

While giving and helping those in need is vital and necessary,
the manner in which we do it is important. Are we fostering
community, or reinforcing our domination system of class
structure? Robert Kennedy while passionate about helping the
poor felt that handouts to the poor created problems, "welfare
has done much to divide people, to alienate us from each
other."[19]

I will never forget attending a Sojourner's conference in
Washington in January of 2000 and hearing a street minister from
Boston tell how his rich benefactors balked at meeting the youth
that their donations were helping. Mother Theresa lived with and
worked for the poor. Jesus would hug lepers.

My Catholic Worker friend Paul Frazier always reminds me
that we should strive for community over charity. Dorothy Day
says that it was Peter Maurin's vision to start the Catholic Worker
so that, "a new social order wherein man is human to man and
which can be built up on the foundation of the works of mercy
and voluntary poverty."[20]

Day speaks of the joy and fulfillment brought about by the
community of sharing and giving: "Sometimes in thinking and
wondering of God's goodness to me, I have thought that it was

because I gave away an onion. Because I sincerely loved His poor, He taught me to know Him. And when I think of the little I have ever did, I am filled with hope and love for all those others devoted to the cause of social justice."[21]

The actions and the words of the prophets spoke to God's love for the poor, the pariah, the imprisoned and the persecuted. As Rich Sider in *Rich Christians* notes; "God not only acts in history to liberate the poor, but in a mysterious way that we can only partly fathom, the Sovereign of the universe identifies with the weak and the destitute."[22]

There is a mystical and spiritual power that comes from helping and embracing the poor and the pariah. Help and hug those in need and suffering, it will be spiritually transformative.

The Force of Unity

Unity is a force in the universe that looks to bring us together and in doing so gives strength to the unity that it creates. Unity tugs at our soul, we long to reconnect with our higher self the God within us, to those around us, to Mother Earth. According to Tantric philosophy, Shakti (energy) must reunite with Shiva (consciousness) to move us closer to God. The Native American prophet the Peacemaker showed us the power of unity when he brought together disparate and warring people to create the Haudenosaunee. The ultimate result was peace.

Spiritual exercises done with others, meditation, sweat lodge or group pranayama (breathing), have the power to brandish attachments, strengthen us spiritually and help us connect with the divine. Unity creates a synergy far greater than the sum of its parts. Jesus tells us, "Again, truly I tell you, if two of you agree on earth about anything you ask, it will be done for you by my Father in heaven. For where two or three are gathered in my name, I am there among them."[23]

Group spiritual practices create synergy.

To understand the ability and power of unity to transcend the physical plane we need to realize that beneath the delusion of reality is God and by uniting we drop, however miniscule, part of our false identity of self. All is Brahman, all is God. So when we gather together and unite in spirit and faith we are connecting with God and reversing the process of separation and fragmentation. Whether we gather in a community of faith or do group spiritual exercises, or gather to bear witness, our spiritual experience is enhanced. God becomes a powerful presence.

The Isa Upanishads teaches us that seeing oneness of everything brings a profound sense of unity: "And he who beholds all beings in the Self, and the Self in all beings, he never turns away from it. When to a man who understands, the Self has become all things, what sorrow, what trouble can there be to him who once beheld that unity?"[24]

Dr. Wayne Dwyer tells us that we should never exclude others because in doing so we elevate ourselves and lose our connection to Source.[25]

Unity and the coming together of all people's, races, creeds, sexes and ages is a dominate tenet in the Baha'i faith. The Bahaullah tells us, "The fundamental purpose animating the Faith of God and His Religion is to safeguard the interests and promote the unity of the human race, and to foster the spirit of love and fellowship amongst men."[26]

Increasingly over the last millennia various faiths and prophets like the Bahaullah have shown us that God's plan is for us to unite. The more that we can come together the more we advance the project. Put your faith in God and take inspiration from the Quakers: "Friends have always held dear the belief that the Light would bring them into unity."[27]

Love is unity.

Chapter 12

Sacred Earth—Creating Sacred Space

We are integrally connected to Mother Earth. She gives us sustenance, maintains our health and well being and helps with our
spiritual development. Without her we are doomed. All of the
tell tale signs of disease and spiritual decay are becoming
increasingly apparent, global warming, storms and environmental upheaval. If we do not change and work with Mother
Earth there will be dramatic declines in the human population as
so often has been the case in history.

The solution to healing Mother Earth and ending global
warming and pollution resides in our hearts, not in our minds.
Relying on science, technology, and free market capitalism is
what got us into this mess; their solutions will only exacerbate
the problem. Mother Earth needs love, mercy and compassion
for herself and we for each other.

Harmony with all living beings and Mother Earth is created
by living in balance with Mother Earth. By understanding how
Mother Earth's vital parts work and then focusing our efforts on
strengthening those vital parts we can exponentially increase our
spiritual efforts to transform the world.

This is a very radically different approach to healing Mother
Earth and bringing about our transformation. It relies upon the
heart. It also imitates the Tantric and Vedanta spiritual exercises
of focused meditations on the vital parts of our subtle body such
as the chakras, that have shown to hasten one's spiritual
evolution. For example, focusing on the ajna chakra (forehead
area) can increase our intuitive ability. Here we will be not so
much focusing our meditations on Mother Earth's vital parts, but
meditating at them. The components of Mother Earth's subtle

body such as spirit lines will be emphasized over picturesque places, or creating a formation to take in the sun.

Knowledge of Mother Earth combined with loving intentions and a little prayer can be incredibly healing.

Sacred Manner

Make an effort to be aware of your spiritual footprints (thoughts and actions), both good and bad. Thinking and doing things in a sacred manner will help reduce the mental pollution you create with negative thoughts. Native Americans talk about doing things in a 'sacred manner,' particularly ceremonies and efforts to connect with the divine. In a sacred manner means that your intention is focused in a way that gives honor, respect and love. Earlier we heard from Sun Bear tell how when earth changes happen people will move in a sacred manner in Chapter 3 "Sweet Mother Earth." Similarly Black Elk in *Black Elk Speaks* and *The Sacred Pipe* talks about doing things in a sacred manner. In preparing a purification lodge as part of lamenting for a vision, Black Elk says, "In a sacred manner you must also gather the rocks and sage"[1]

Black Elk tells us that sacred manner includes blessing and honoring that which one takes to create the purification lodge: "First you should build and Inipi lodge in which we shall purify ourselves, and for this you must select twelve or sixteen small willows. But before you cut the willows remember to take to them a tobacco offering: and as you stand before them you should say: 'There are many kinds of trees, but it is you whom I have chosen to help me. I shall take you, but in your place there will be others.'"[2]

The act of honoring the willow tree that has been cut down helps reduce the consequences of destruction by counterbalancing it with a positive thought and action. It is a far cry from our modern day world where it is all take, cut, burn, excavate and various other forms of destruction without thought. We should always try to avoid the negative, but if something happens we should counter it with a positive thought, or prayer.

A further glimpse of sacred manner comes from Black Elk when talking about how we should always keep in mind and honor Great Spirit and Mother Earth: "It is the Earth, your Grandmother and Mother, and it is where you will live and increase. This Earth which He has given to you is red, and the two-leggeds who live upon the Earth are red; and the Great Spirit has also given to you a red day and a red road.[3] All of this is sacred and so do not forget! Every dawn as it comes is a holy event."[4]

For example, when the Tree of Peace (White Pine Tree) we had planted on the shores of Onondaga Lake (Syracuse, NY) died and needed to be replaced, my Mohawk friend John Sardella said that the process should be done in a sacred manner. I removed the dead tree and brought it to his home. A sacred sweat lodge was held honoring the tree and using it to fuel the fire that heated the rocks. Then there was a period of grieving before the new tree was planted. Finally John led the ceremony for planting the new Tree of Peace.

Doing things in a sacred manner means applying thought to what we do and think. It is about honoring God and Mother Earth and all of creation. When we do things in a sacred manner we give greater strength to our thoughts and actions.

Self Cleansing and Purification

You should develop a regular routine of cleansing your subtle body. We are continually coming into contact with a variety of unseen things that can cause harm. These unseen things can attach to you and cause problems if you do not remove them. Cleansing your subtle body will help remove negative thought forms before they make a home in your subtle body.

Cleansing begins with meditative visualizations of seeing white light coming into you and clearing your aura of any negative thought forms or attachments you may have picked up. At the same time you should also physically clean your aura to try and remove any psychic rubble. The idea is to reinforce the intention with physical action. There are a variety of techniques you can use.

Dust yourself off. Just as you would remove dust or lint from a jacket or sweater, similarly imagine flicking debris off of your aura 6 to 24 inches from your body. As you are brushing your aura see attachments and negative thought forms being expelled.

Wash your subtle body. Imagine washing your aura as you would your body in the shower, keep your hand movements confined to an area 6 to 24 inches from your physical body.

Washboard technique of cleaning your aura. Visualize that you have an imaginary washboard attached to your aura. Go up and down with your hands on the washboard as if you are scrubbing your clothes. Move the washboard around to various parts of your aura.

Smudging is a wonderful way to clean your subtle body by brushing it with an object. You can smudge yourself or have someone else smudge you. Decide whether you will be using a burning smudge stick (sage, rosemary, sweet grass, and incense) or an object such a feather. Next wave the smudge stick around your subtle body, about to 1 to 3 feet from your physical body from head to toe. Make sure to get your back side as well as your front side.

During the process of smudging visualize yourself being cleansed of any negative thoughts and attachments. Use a prayer or a mantra such as 'clean and protect' while you are smudging. Give thanks when you are finished.

As you progress with your smudging and develop your subtle body, the feeling of being smudged will become very pleasant. You will get the same sort of relief that accompanies scratching part of your body that has an itch, only it will also make you feel spiritually better.

All the techniques used to cleanse yourself, if done regularly, will help develop your subtle body and make it progressively easier for you to remove negative thoughts and attachments. The physical process combined with the mental focus can become very powerful.

Asking for protection is always a good precautionary measure. However, developing your ability to protect yourself takes time and effort.

Once you have developed sentience of your subtle body you may want to start healing areas damaged from negative consciousness by "laying hands" on that area of your subtle body, or through some other technique such as therapeutic touch or reiki.

Creating Sacred Space

While all space is special and part of God, not all of it is positive. Sacred space is a space filled with love and positive intentions. It has a positive geographic samskara and makes you feel good. Sacred space can help raise your consciousness, cleanse you, heal you, and help you connect with the divine, or rid you of negativity. Any space can be transformed into sacred space.

When we create sacred space we provide a loving and healing space for others. We enable Mother Earth to better provide us with her love and beneficial energies. We may also be removing blockages in the flow of prana created from bad consciousness. Creating sacred space means making a positive geographic samskara at a particular location. Over time it can draw energy features such as a vortex. Ultimately creating sacred space is all about recognizing Mother Earth and working with her to advance our collective spiritual evolution.

It is vital to have sacred space to counteract the negative space we traverse through during the day and purify our subtle body. Part of our essence remains as we pass through a particular location and we pick up part of what we travel through. How much is a function of our intention and the time we spend at a space. If we are constantly living/walking in a sea of negativity we will be swept away by the tide. Remember samskaras are the biggest impediment to our spiritual evolution.

Wherever we pray or mediate can become sacred space. The process of mediation reduces negative geographic samskaras. The spiritual strength of the meditator, the time spent there and the spiritual condition of the location will determine the potency applied to creating sacred space. One of the reasons we should meditate at the same place is to create a positive samskara there. We also create a samskara that increases the propensity and ability to meditate. This is one of the reasons we are told to meditate at the same place. I call these thought forms 'meditation samskaras'. Over time a meditation samskara can create a vortex of cosmic prana that will significantly increase the amount of cosmic prana drawn to where you meditate and help raise your consciousness.

Sacred spaces are bright beacons of God's light and divinity upon Mother Earth. They instill the positive in all who come in contact with them. Creating sacred space is vital if we are going to heal the world and Mother Earth.

Native American Ceremonial Sites
Native Americans have a wonderful custom of creating ceremonial or prayer sites wherever they go. While traveling they would stop along a trail or at the intersection of several trails to pray, do ritual or say thanks to Mother Earth. Over time these places would become divine places, especially if the site was close to some special feature of Mother Earth.

Many of the places of worship in North America and in other places around the world rest on old ceremonial or sacred sites. We are naturally drawn to the positive geographic samskaras of the place whether we know it or not. The spiritual embers from Native American ceremonies can remain for a long time. When I look for sacred sites I often find that the places I select were previously ceremonial sites. I have also found that certain geographical areas can be dotted with numerous places of worship such as in Ithaca, NY[5] where several Native American trails intersected and people would spend time in prayer there.

Follow the spirit of the Native Americans and make a ceremonial or prayer site somewhere outside of your home. Choose a place that you frequent regularly, on the way to work, when walking with your dog or someplace you jog. It could be a section of a subway station, a bus stop or a busy street corner. Make an effort to try and be at that exact location on a regular basis. It does not matter where it is, or when you go there, only that you will go there on a regular basis for a while.

The location does not have to be formalized. You are the only one that needs to know where it is. Go there and pray or meditate there for a few minutes every few days. You don't have to sit down, you can stand. You don't even need to close your eyes. In fact, you can do it from a distance. All you need to do is visualize the specific space being cleaned. Just make the effort and have the intention of creating sacred space. For example, while meditating at home you might visualize yourself being there and divine light shining upon the space.

Welcome circles

I like to focus my prayers and intentions for specific purposes when creating sacred space. One of my favorites is to construct a welcome circle to greet people. I often create welcome circles at sacred sites to act like the foyer or vestibule of a house. Like a vestibule, a welcome circle is the introduction to a place. It is the place where you check your coat, are greeted and prepare to enter the home. The welcome circle is similarly meant to prepare you for your upcoming spiritual experience.

In constructing a welcome circle my first intention is to cleanse prospective visitors of any negativity. I say a prayer and meditate on people being cleansed in the circle. I reinforce this with the action of visualizing white light cleaning me every time I am in the space, or when others are in the space. I further build upon the samskara through the physical action of smudging. If you cannot smudge at a location go somewhere you can and visualize doing it at the welcome circle.

Secondly, you want to welcome someone. I ask God to please help those in the welcome circle help connect with their higher self and Source. You are trying to welcome them to the divine. The physical manifestation of this is the hug. Give everyone a warm hug with heartfelt emotion that enters the circle.

Finally, I ask the Creator to please help them to learn from this experience and take that knowledge home. I physically reinforce this later by giving thanks for all that we have gotten from the experience.

You can create a welcome circle, a healing circle or whatever type of circle you want. It is not necessary to formalize a sacred space with stones, although it helps. The key is effort and your intention.

Stone Circles

Circles are one of the oldest formalized sacred structures in the world. At the higher planes of existence a circle represents, or is the manifestation of, unity and oneness. It symbolizes the coming together that is part of God's plan. The circle can also be used to represent the downward arc for those that come together for power, control and exclusivity.

The circle enhances our spiritual experience and development. If we draw, or cast, a circle of stones with the right intentions and purpose we can create a very loving place. Feng Shui tells us that one of the most auspicious places is where the wind meets the water because the chi (prana/energy) will be retained there. A circle of stones can be such a place.

If we pray, meditate, do ceremony with the intention of love and the divine within the circle we will be creating a positive geographic samskara. The stones will help retain that consciousness; it will also help form a positive (clockwise) movement of cosmic prana. Over time this movement can turn into a natural vortex of cosmic prana. Spiritual exercises done in a stone circle with a natural vortex will give you a potent source of cosmic prana to help raise your consciousness.

Interestingly the stones making the circle block earth prana(chi, the life force) and other lower forms of prana from entering the immediate area above the stone circle. Prana will be flow around the circle once it comes in contact with the stones and not enter the area within the circle.

Sang Hae Lee in his doctorial dissertation on Feng Shui notes says that we should look to build homes where chi (prana) accumulates: "Human dwellings should be selected where the earth is least hurt with consideration for the surrounding conditions of nature. On its cosmological and metaphysical level, feng-shui specifies that the site for human habitation should be selected where cosmic chi accumulates, because this is the place where "yin and yang" join and reconcile, and the heaven and earth accumulate." [6] A stone circle is such a place.

Labyrinths and Other Structures

There are many more types of structures besides a circle that one can create. In creating these structures I recommend that the physical structure and any related actions be reflective of its intention and purpose. It should be like the ritual of the sweat lodge where the process of physical sweating reinforces the spiritual cleansing and purging.

For example, consider a labyrinth. If the intention of the labyrinth is spiritual, to connect with God, get insight and the like, then its physical makeup and movement should reflect looking within. Most labyrinths do this through their circular form and movement of labyrinth walkers from the outside in.

The difference between various labyrinths is the path that you will walk. Most have you twisting and turning in different directions. Others are like a maze the will have you thinking about where to turn next. While moving in different directions may be reflective of the twists and turns in life and may help you focus away from the outside world, does this help you in your spiritual journey? Longer term, a twisting and turning course with numerous changes of direction within a labyrinth can create blockages that impede the proper flow of prana.

Your movement in a labyrinth should be reflective of your spiritual quest. You are looking to reinforce your spiritual pilgrimage within and your movements in the labyrinth should be a steady progressive movement in a clockwise movement. A spiraling movement inwards, without twists and turns, gives strength to your progressive and steady movement within.

On the pranic plane your steady clockwise direction within reinforces a positive clockwise movement of prana. Remember that your spiritual body leaves a trail of consciousness and energy. A labyrinth is one way of taking advantage of this. Over time the steady clockwise movement of labyrinth walkers increases the chances of creating a natural vortex.

The contemplative nature of the labyrinth walker creates a positive geographic samskara that will assist others that follow later in getting answers or connecting with God. I have received many insights in labyrinths thanks to the efforts of those that have walked in them before me. For example, the insight about using the movement of labyrinth walkers to create a natural vortex of prana was revealed to me while walking the labyrinth at the Foundation of Light in Ithaca, New York.[7]

Stone Talismans

I work a lot with stones to create a variety of structures to tap into and enhance Mother Earth. Stones are a great natural tool. I would recommend caution in dealing with any manmade gizmo's that promise to clean or enhance space.

Stones have certain qualities, and as noted earlier they can block the flow of earth prana and other lower forms of prana. They are also a wonderful talisman that retains the consciousness of what has occurred around them. They can be used to facilitate the process of creating sacred space.

I generally meditate on stones when creating sacred space outdoors. The stones better help retain the samskaras from my meditation and subsequently help in drawing cosmic prana to create a natural vortex. The stones can be of any size and shape. They will also retain much of their vibe if they are transported to another area. I refer to a stone charged with spiritual intentions as a 'prayer stone' and they are a great tool to use for bringing light into very dark areas.

In creating a stone circle or any other stone structure, you might want to spiritually clean your stones before you use them. Since stones carry the samskaras of previous experiences they may have been injured in a fire, or ripped from the earth, or been where an animal was killed. You clean a stone by holding it and saying a prayer. Your intention is the key.

Cleaning Space—Bringing Light into Darkness

Where ever we pray, mediate, do good acts, work for social justice or have good thoughts, we are creating or adding to sacred space. If the space is filled with negative thought forms then we are beginning the cleansing process.

While much of the world is covered with negative geographic samskaras there are some places that are very negative. Places of violence, war, slavery and where horrific things have happened, or where Mother Earth has been violated through mining and deforestation are places that need to be cleansed because they will continue to attract the same sort of bad behavior. They are also a sore on Mother Earth's body that will disrupt the smooth functioning of her subtle body through such things as altering or blocking the flow of prana.

Bringing light to darkness is a vital, but potentially a dangerous mission. Very vile locations can be frequented by entities of all sorts. When you attempt to cleanse a space you may be potentially taking them on, invading their turf per se. You must be extra careful. Being in the company of others and with someone that is experienced in clearing space is helpful.

The most basic way to cleanse a space is to pray or meditate there. Meditation in particular will help clean up negative geographic samskaras. You can supplement your spiritual cleaning with the action of smudging. Light a smudge stick or use a feather. Wave the smudge stick around in wide swaths. Move your hands around as if you were physically cleaning the space. I move my hands in a clockwise circular motion as if I were cleaning a window. Do this in a sacred manner.

While you are smudging visualize light and God's love coming in. Ask for divine help and guidance. Visualize angels by your side cleaning the space.

If you have a large space to clean, begin working from the periphery inwards. Choose one place at the outside edge and meditate there, preferably a location that is on a spirit line or at the intersection of several spirit lines. Select another location close by moving in a clockwise direction at the periphery and make sure to get all the spirit lines. Meditate there. Progressively move towards the center in a clockwise motion. Clean up small bits of space and take your time.

Always smudge yourself after you have cleaned up a space, either with a smudge stick or with your hands.

If a space is too contaminated with negativity you might want to begin your cleaning at home, or at a sacred space. This way you will not be engulfed and depleted by all the negativity of a space. Imagine the space that you are going to clean and pray for it to be cleaned, ask God to help, or visualize light coming into it. Over time your efforts will bear fruit. You can also create a prayer stone that will blend in with the surroundings and leave it there.

Remember meditation, contemplation and focused prayer is the best ways to clean space. Mediation is what I call the 'heavy lifting' necessary to clean a space. Many people think that smudging or spreading some tobacco can clean a space. This can supplement your effort, but you need to meditate. Remember Buddha found enlightenment through meditation. Because we are like Mother Earth and Mother Earth is like us, it is meditation that will bring enlightenment to her as well. Meditation also draws in beneficial essences from Mother Earth such as cosmic prana. It also plants a very potent seed thought, or samskara, that will look to bear fruit and attract more of the same. Meditate.

After cleaning a space, particularly if it has very negative geographic samskaras, you need to cleanse and heal yourself afterwards. I go to a sacred space that has a great vibe and divine features of Mother Earth. Visiting and meditating at such a place will better help you cleanse yourself and meditating there will replenish you. While cleaning space is vital, you have to be very judicious about it because it can be very draining. Just as you can pull a muscle by trying to lift too much you can similarly injure yourself by tackling a spiritual cleaning that is too big.

Cleaning a space brings numerous benefits such as getting rid of negative samskaras. In the process you will be healing Mother Earth and reestablishing her proper functioning to an area. You will be building a positive geographic samskara that will help others, and over time, you may even create a natural vortex. As you progress with your cleaning more of Mother Earth's mysteries may be revealed to you at a particular location. You may notice that there are subtle features of Mother Earth there that you previously did not know existed. You might attract nature spirits, or angels. Cleaning space can be like a spiritual treasure hunt as anything and everything is possible.

Community clean-up

You should endeavor to spiritually cleanse your community both individually and with others. A serious effort has to be made to counter the negativity and violence that occurs in communities across the world by cleaning them up.

There are little things that you can do that can make a big difference. If when driving a car you see a dead animal or a place where an accident has occurred, you should pray. You should not become so absorbed that you get into an accident yourself, but you can send a little love. I always try to ask God to help an animal killed by a car. I have developed a motion with my hand that has become almost second nature. I do it for dead animals and for crosses I see by the side of the road marking where someone has died in an accident. It immediately invokes a blessing to be sent out. I don't stop, I just send out the prayer.

It is critical for communities to gather in prayer at places where tragedy has occurred especially if someone has been murdered. Visit a shrine that has been set up at the site of a tragedy. A prayer vigil with a group of people at the site of a tragedy can be incredibly healing for a space.

Consider participating in a pro-active group that prays in your local community. Prayer vigils and prayer walks are held in communities around the world. Such efforts are a powerful counter to the negative geographic samskaras that permeate our cities streets. I have been on prayer walks with Revered Larry Ellis and others in the city of Syracuse's south side. They are powerfully healing and transformative as well as a great way to connect with others. I am always amazed at how many people want to pray with us, particularly young men. I also try to join Helen Hudson and others from Mothers Against Gun Violence in Syracuse at prayer vigils which are held where a murder has occurred. Such prayer vigils are vital; they help console victim's families and help the community heal and provide a powerful spiritual force to where something terrible has happened. They bring light to darkness.

Connecting with Mother Earth

Make an effort to see Mother Earth and not the entrapments of our material culture. Focus on nature, not inanimate objects. This will help develop your sentience of Mother Earth and help you to better absorb what she gives to us. Gaining sentience of Mother Earth is a gradual process that will improve your health, reduce your vulnerability to disease, increase your feeling of well being and raise your consciousness. The more we focus on Mother Earth individually the greater the collective benefit, as she will respond by progressively enhancing the environment for us. Reinvigorating this nurturing relationship is critical to our collective evolution.

While Mother Earth is full of beauty one should not make physical attributes the driving motivation of where to pray or meditate. You should endeavor to choose a location that has consciousness in the form of spirit lines, or a place that has energy. Many ancient cultures chose locations with powerful energy formations containing lots of energy lines (nadiis), or earth chakras that regulate earth prana. I am not a big advocate of meditating at energy points, although they do hold merit and bring about several benefits. They can be incredibly invigorating and give you a big rush. I have seen them turned into a natural espresso bar by surrounding them with a circle of stones. A few minutes in one of them are like drinking a few espressos. Earth chakras can help your body learn to absorb energy, which can be very helpful. If you want to get a dose of energy I would suggest a natural vortex of earth prana, not an earth chakra.

I recommend locations that have lots of consciousness in the form of spirit lines, particularly locations where several spirit lines intersect. Spirit lines, lines of consciousness, what many call ley lines, carry consciousness. They emanate from Mother Earth's soul and are meant to carry divine attributes around the world.

Unfortunately they are affected by the negative consciousness from our bad thoughts and actions that they encounter along their path. This not only impedes their ability to carry divine consciousness but they begin to carry the negative consciousness that they encounter along their path.

It is important to pray and meditate on spirit lines or where several spirit lines intersect so that you will be sending out love beyond where you are meditating. How far will vary, but chances are that your intentions will carry well beyond the point where you are praying. Meditating at the intersection of spirit lines will exponentially increase you affect by greatly increasing the space that you will be influencing. Spirit lines that carry negative consciousness can also create blockages to the proper functioning of Mother Earth's subtle body. When one meditates, or prays on fouled spirit lines, you will be helping clean them and healing Mother Earth.

Finding spirit lines will take some work. You might want to learn how to dowse. You can go to your library and get a book on dowsing, or search websites online. In the Recommended Reading list in the back of the book you will find several helpful links on dowsing. Most dowsers hold regular meetings at which they teach the basics of dowsing. Go to one of these meetings with a pair of L-rods (brass dowsing rods) and asked to be shown a ley line. With a little practice you should be able to learn how to find them.

You can also try to intuitively find spirit lines through contemplative meditation. You can also ask a dowser to assist you.

To be better able to connect with Mother Earth you will have to learn how to sense her presence. Try to sense consciousness, whether it is spirit lines or geographic samskaras. You will feel consciousness in your chidakasha, the seat of your consciousness or mind-screen, in the greater part of your forehead area. Focus on this area when you meditate on spirit lines, or a place with a very good vibe (positive geographic samskara). Some people feel consciousness in their subtle body, or tingling in their hands or legs, or their legs buckle, or the place just makes them feel good.

Similarly go to a highly charged energy area and try to feel the energy in your ajna chakra, or third eye. The sensation should be more localized to your ajna chakra than your whole forehead area. You may also feel a tingling sensation coming up your legs, or around your aura. The same sort of thing can happen with sensing consciousness. Or you might feel invigorated and charged up. My dogs always get very hyper when they are in a field of energy vortexes.

What ever sensations come to you try and build on them. Give thought to them and in doing so they will get stronger. Meditating at a place and focusing on the sensations will help you develop sentience. If you don't feel anything imagine feeling something, or visualize white light coming into you and try to feel it. Try to learn to feel the difference between consciousness and energy.

The Energy Conundrum
In connecting with Mother Earth it is important to understand the difference between energy and consciousness. The term energy has a lot of meanings. The most basic definition of energy is prana, or chi, the life force that sustains us. Its most rudimentary manifestation in the physical plane is the material world of objects. The physical objects that we see around us, including our body are made of prakriti, or some form of material energy. The world per se is not pure energy. It is made up of combinations of energy (prakriti) and consciousness (purusha).

People frequently use the term 'energy' to describe a place or an event. This is not always accurate because while there may be a lot of energy somewhere there could also be a lot of consciousness there as well. Remember everything is Brahman, so everything has a bit of consciousness. In fact it is the consciousness of others, the thoughts forms and samskaras that are setting the tone in the world. We really don't have a word to describe the consciousness of a place (geographic samskara), or the combination of energy and consciousness of a place. That could be because our approach has been on energy.

It is difficult to say why we have focused so much on energy. It could be that in our progressive plunge into the material world we have become increasingly focused on material objects and hence energy. It could be that geomancy techniques such as feng shui have focused solely on energy and ignored consciousness. It could be that the Tantric and other traditions concentration on energy have contributed to this focus. It could be that our thinking has been influenced by science which does not believe that the universe has consciousness and that God exists. Whatever the reason, our focus on energy has given us a myopic view of the world.

Not only is the exclusive focus on energy wrong, it is dangerous. The focus on energy reduces our attention to consciousness, spirit and God. Are you made up of energy, or consciousness? Both, but which is more important?

Energy, or prana, is regulated by the pranic plane, which is the plane of nature spirits and other entities. When we focus exclusively on energy, either through spiritual exercises or where we pray and meditate we can attract entities that are looking to feed on this energy. Energy is important and vital, but you have to be very careful how you work with it.

Focus on Consciousness
It is our bad thinking or negative consciousness that generally causes disease within us. Similarly it is our negative consciousness that causes diseases for Mother Earth in the form of hurricanes, violent storms and earthquakes. Because of that, I think it is important to focus on our consciousness in healing Mother Earth by praying at spirit lines. The world is covered with

bad consciousness that needs to be cleansed.

While energy can give us sustenance, it needs consciousness to do its work as Swami Niranjanananda Saraswati tells us:

> Purusha (consciousness) must always work in cooperation and union with Prakriti(energy). Without prana(energy), consciousness is unable to create. There must be an underlying force which is transformed into various objects and forms. On the higher level of experience, prana and consciousness are one. On the mundane level of existence, however, they are mutually related and interact with each other. They are, in fact, mutually dependent entities, at times merging and at times becoming separate. Prana can thus be affected by consciousness and vice versa.[8]

Energy, or prana, is necessary to help elevate our consciousness but should not be the main focus of our efforts. I don't think that we should follow the path of the ancient cultures that focused their efforts in creating ceremonial sites by earth chakras.[9]

Energy work is important. As was noted earlier over time Mother Earth responds to our good thoughts and prayers by sending her equivalent of kundalini energy to where we pray. This energy can significantly expand the consciousness of a place and create a spiritually enthralling location. The only caveat is that you need to be careful in how you work with energy and not make achieving it the ultimate goal of your practice.

Spiritual Ghettos

The unfortunate thing is that the bulk of the world is covered with negative geographic samskaras that hamper our spiritual evolution and encourage us to repeat the bad behavior that has gone on before. These are places where we have injured Mother Earth's subtle body preventing it from properly functioning. Not only is the world covered with negative geographic samskaras, but some of the most visually spectacular places in the world have the worst geographic samskaras. That is because those places have been coveted and fought over for millennia. The water side location or the mountain top with a view are places where people have literally died for.

Ravaged crime invested inner city communities are called ghettoes and are places where the samskaras of violence, domination and hopelessness can be found. There are also roots of faith within these communities where deeply spiritual people reside who serve their communities and the greater good. Conversely some of the most perniciousness consciousness lurks in affluent suburbs and cityscapes and are the equivalent of spiritual ghettoes. Built on white flight, wealth and power, they often have a very negative vibe.

The message is clear, don't look with the eyes, but try to feel the consciousness of an area. If you do look, look to avoid the markers of wealth and its ostentatious trappings. Avoid places that have signs of negative consciousness, such as posted signs, walled communities or other signs of exclusion, privilege and domination. You should avoid such places, or judicially visit them realizing that they are going to diminish your subtle body and consciousness.

Turning Your Home into a sanctuary

You can and should transform your home into a sanctuary to get away from the reach of negative geographic samskaras. You need to spend as much time as possible in a place of light with a positive consciousness. You need a place to heal after passing through the negative geographic samskaras that blanket the world and have a refuge after you have been working to create sacred space where darkness once was. A home sanctuary will help with your spiritual development because the positive geographic samskaras of love and spiritual growth will help elevate you. The natural vortices of cosmic prana and other features that develop over time help cleanse and heal and raise consciousness.

The task of healing the world is overwhelming since it is covered with negative consciousness and talismans of all sorts. You must judiciously choose the places you want to heal and turn into sacred space. You cannot clean all of your community, nor can you advance if you are constantly surrounded by negative geographic samskaras as they will inhibit your efforts. Your home needs to be your sanctuary, a place where you will be protected and can replenish yourself.

Swami Vivekananda believed that we should have a room dedicated solely to spiritual practices if possible. He said such a room should have flowers and pleasant pictures, and that incense should be burned there in the morning and in the night: "Have no quarrel or anger or unholy thought in that room. Only allow those persons to enter it who are of the same thought as you. Then gradually there will be an atmosphere of holiness in the room, so that when you are miserable, sorrowful, or doubtful, or when your mind is disturbed, if you then enter the room you will feel peace…The fact is that by preserving spiritual vibrations in a place you make it holy."[10] He notes that the original idea behind the temple and church was to make a holy space, but in most instances that has been lost.

While Swami Vivekananda suggests making one room a sanctuary, which we will talk about shortly, we can turn a whole house into a spiritually positive place. The process of turning your home into a sanctuary begins with the decision to do so. Meditate on how you are going to accomplish this and visualize what you want it to be. Decide what spiritual attributes you want your home to have such as peace and serenity, welcomeness or the divine. Consecrate the decision through the action of a ritual of some sorts, smudging, an offering, burning incense and a prayer.

Formalize your decision by making an altar in your home.

The altar can be as simple as a stand with sacred texts in one part of a room. You should do this along with your other efforts in transforming your home in a sacred manner. The altar should be a dedicated area where you routinely pray and meditate within the larger sanctuary of your home. Your entire home needs to be sacred.

Focus on consciousness in the transformation. Try to situate your altar near a spirit line, preferably at the intersection of several spirit lines. When we mediate in the same place regularly we create a 'meditation samskara', a thought form that will assist with meditation. While I have one central location for prayer and meditation where my altar is, I periodically meditate in different places. I also do certain types of meditations at certain places. For example, I do a mind emptying or relaxing meditation in my living room while my concentrative meditation and other spiritual exercises are done by my altar.

You need to get the consciousness of your home focused on spirit, questing and God. Look at the talismans in your house. Get rid of anything associated with violence like guns, hunting knifes, violent DVD's/CD's/Books. Get rid of ostentatious symbols of wealth and domination, expensive furniture/jewelry, items associated with the stock market and other symbols or wealth. Try to get rid of things that represent or carry negative, demeaning consciousness. Make your home as simple as possible and reduce electronics to the bare necessities. Bring sacred texts and other artifacts that remind you of God and spirituality into your home.

Physical cleaning will help with your spiritual cleaning, through the physical action of cleaning and the generic character of the thought form of cleaning; but do not go overboard and become obsessive.

Spiritually clean your home by smudging it on a regular basis. The physical ritual of cleaning will help facilitate purging your home of negativity.

Begin the spiritual cleaning of your house from the outside in by meditating at the periphery of your property. You are trying reinforce the action of looking within. Chose a starting point to begin the mediation process of creating a positive vibe and meditate there. In a gradual clockwise motion over the next few months progressively move your meditation locations in a clockwise motion around periphery of your property. Make sure to meditate on all spirit lines coming into your home. When you get close to where you started from move inwards towards the house as if you were creating an inward moving spiral.

Create welcome circles at the all of the entrances of your home. Not only should you pray and meditate in these areas but you should try to pause for a brief prayer (if only momentarily) whenever you pass through one of the welcome circles. Make an effort to greet guests there. Periodically clean your aura there and meditate there.

Get together with a bunch of friends for group meditations. Take turns meditating at each others homes. Group meditation creates a synergy that will significantly increase the benefits of meditating.

There is a lot more you can do to transform your home into a sanctuary. I have a prayer bench in my front yard that sits on an old ceremonial circle where everyone is welcome to pray. I try to pray and give thanks there with friends when they come by. You could put a labyrinth in your backyard, a stone circle, incense burner and a lot more. Decorate your home with spiritual accou-trements such as candles, incense burners, chimes and pictures of sacred sites/prophets/people showing love/unity. You can make holy water by praying over the water and then using it to spray your home or your garden. You can leave offerings in your backyard. I have also turned my backyard into a Spirit Keepers[11] site with large stones marking places in the field of consciousness. There is a lot more you can do.

Honoring Mother Earth

Giving thanks to Mother Earth is a great way of honoring her and giving her recognition. Such recognition is an important counter to the abuse and disregard she has had to suffer because of callous disregard.

As the head of the Tree of Peace society Chief Jake Swamp has planted many Trees of Peace around the world. In addition to telling the story of the Peacemaker he talks about giving thanks to Mother Earth. In his book *Giving Thanks — A Native American Good Morning Message* he says:

> To be human is an honor, and we offer thanksgiving for all the gifts of life. Mother earth, we thank you for giving us everything we need. Thanks, deep blue waters around Mother Earth, for you are the force that takes thirst away from all living things. .We give thanks to green grasses that feel so good against our bare feet, for the cool beauty you bring to Mother Earth's floor... And most of all, thank you, Great Spirit, for giving us all these wonderful gifts, so we will be happy and healthy every day of our lives.[12]

Another way we can honor and give strength to Mother Earth is by praying at sacred sites. With sacred sites here being defined as places that are vital to Mother' Earth's makeup, or where great spiritual or social justice events have transpired. You should look for places that have simplicity in structure and avoid gargantuan or ornate structures as they may be reflective of the ancient equivalent of our modern mall. Also note if the location is drawing all sorts of carpet baggers and peddlers of spiritual merchandise or services. This would indicate decay and negative geographic samskaras. It could be a nice place but needs to be cleaned. Jesus teaches us that you will know the true prophet from the false prophet by the fruit that they bear.[13] It is the same for the land. Look at what has transpired at a place to help you gauge it.

When you visit a sacred site you will be reinforcing divine attributes there and making them stronger. You will also be will be spiritually gaining from them. Our collective thoughts are like building blocks as Thich Nhat Hanh notes, "By taking refuge in a spiritual community, you become rooted in the experience and, wisdom and consciousness of those who have gone before you. At the same time, you yourself become a perennial root for the flowering of all the young people who will continue after you."[14]

A pilgrimage, whether close to home or far away, can be spiritually transformative, particularly if you go to a place that has developed powerful geographic samskaras from all its visitors and more so if it has some divine attributes of Mother Earth. For example, the chapel Kateri Tekakwitha[15] in Fonda, NY, currently a Roman Catholic shrine, can bring you to tears. The chapel sits on a precious piece of Mother Earth and was a place of worship for several Native American cultures before it became a Catholic shrine. The journey of the pilgrimage itself will give you a one pointed focus, even if for only a few hours or days. This focus will enhance your spiritual connection. The physical action of traveling will reinforce the intention of seeking and connecting. While there chances are that you will raise your consciousness and improve your spiritual self much more with the additional boost from the sacred space. You will also be adding love to where you go.

Understand that it is the prayers and meditations of visitors like you that over time have helped make a pilgrimage site divine. If you go to a place that has divine or special aspects of Mother Earth you will be tapping into them at the source. This will give you a very potent dose and help improve your sentience of Mother Earth.

Mother Earth has a variety of special places where we can access her divinity, heal, gain insight or get epiphanies. Try to find such places near your home and turn them into sacred sites through prayer and meditation. The emphasis is on finding naturally occurring divine places and then adding to the love.

Pray at Mother Earth's Soul

I ask that you please visit and pray at what I call Mother Earth's soul[16] in upstate NY, for your personal development and to help Mother Earth heal. The fields of consciousness located within the Greater Central New York (CNY) Area have had a profound influence in shaping America and the world. It was here that the Native American prophet, the Peacemaker, came and gave the Haudenosaunee the Great Law of Peace that would later be the model for American democracy. It was home to Americas Second Great Awakening, the Evangelical movement, the New Age movement and religions such as Mormonism. It was called the 'Burnt-Over District' for all the religions born there because it seemed that what was born one day was soon 'burnt over' by some new spiritual wildfire.

CNY was also the birthplace of social reform and justice. The women's movement was born in Seneca Falls and its greatest leaders, Elizabeth Cady Stanton, Matilda Joslyn Gage, and Susan B. Anthony, lived in the area. It was a hotbed of abolitionism and one of the major routes of the Underground Railroad. It was called 'North Star country' because slaves followed the North Star to freedom here. It was the home of Fredrick Douglass, Harriet Tubman and others. Gerritt Smith, who the NY Times called one of the greatest reformers of the nineteenth century,[17] lived in CNY. What has sprouted spiritually from the greater upstate NY area has shaped America and the world.

The hub of reform and spirituality in the heart of Mother Earth's soul has been called America's 'Psychic Highway'[18] and it is the area between Utica, NY and Rochester, NY. This is where there is the greatest number of fields of consciousness can be found. It is also where we find the greatest number of fields that are stacked upon each other, making the consciousness emanating from there that much more powerful. Fields appear to stretch into Massachusetts and other parts of New England and into Ohio and southern Ontario in Canada.

Fields of consciousness are like the furnace that heats your home, only they warm the world with divine attributes. They are made up of places of consciousness that are like heating coils and are powered by energy lines (nadiis); they radiate consciousness—love, compassion, connection to the divine, self sacrifice and the like. Spirit lines emanating nearby pass over places of consciousness and begin the process of carrying divine attributes around the world.

Fields of consciousness are meant to radiate around the world, but history demonstrates that those living in the closest proximity to the fields are the most influenced by them. While it is difficult to say what field influenced what, some startling results appear when we look at historical events in relation to their surroundings.

Unfortunately many of the fields are covered with thick negative geographic samskaras. Onondaga Lake in Syracuse where the Peacemaker planted the tree of peace and gave us the Great Law of Peace is one of the most polluted lakes in the world. Not only does this pollution prevent Mother Earth from sending out her love, the thousands of spirit lines emanating from there are carrying out the consciousness of toxic pollution. In other words the consciousness of polluting and desecrating Mother Earth, at its vilest level, is being broadcast globally from Onondaga Lake.

By praying around Mother Earth's soul you can help clean the geographic samskaras that have spirit lines radiating the toxic consciousness of pollution. There are other places where Mother Earth's soul is not covered with negative consciousness that will help you expand your consciousness and spiritual self. If they motivated people to greatness in the past they can do so again.

Go to www.MotherEarthPrayers.org   to see the listing of

places. All of the places listed have a great vibe and numerous fields of consciousness. Several locations have been established at Onondaga Lake and other locations that need love and healing. It is important to build up places rather than tackling negativity head on. The places listed will help you sense consciousness and make you feel good, very good.

Cleaning Mother Earth's soul can exponentially help her and humankind. The spirit lines emanating from there travel all over the world. The act of cleansing Mother Earth's soul will become like a giant megaphone sending out love and healing to all corners of the world. The consciousness that once drove people to greatness can begin sending that consciousness out again. Come and visit.

As you begin to appreciate and connect with Mother Earth a whole new world will be opened up to you. Your focus and perspective will shift from appreciating physical things to sensing consciousness, spirit and energies with your sixth sense. Places that you once thought were not so wonderful may now be nice and vice versa.

As you gain sentience your well being and health will improve. You will start to experience a spiritual high what some mystics call bliss. That joyous and wonderful feeling will increase as you get closer to Mother Earth.

Chapter 13

Heaven on Earth

Making heaven upon earth is about freeing our consciousness trapped in the physical plane, by believing, seeing and living in the mystical world around us. It is about visualizing and dreaming of God's plan unfolding before our eyes. It is believing in the world that Isaiah foresaw where, "The wolf and the lamb shall feed together, the lion shall eat straw like the ox; but the serpent—its food shall be dust!"[1] It is about seeing happiness, love and kindness filling the world and the hearts of everyone and everything. It is believing in a world of plenty where no one is in want. It is about being with God.

It starts with a prayer. Focusing, dreaming and meditating of heaven on earth will strengthen the vision. We need to begin to focus on the spiritual plane and renounce the strictly physical world through thought and action. We must become like the ascetic who detaches themselves from the world of sense objects and finds sanctuary with God within and with their spiritual community of friends. Living in the spiritual world will give strength to the spiritual planes while detachment will reduce the sway of the physical plane.

Trapped in the Physical Plane

Heaven on earth is not only a very real possible it is part of God's divine plan. As we noted in the Introduction Jesus tells us in the Lord's Prayer, "Thy kingdom come. Thy will be done, on earth as it is in heaven."[2] Meaning we are called to make heaven upon earth.

The problem is that our consciousness is trapped in the material world. The Srimad Bhagavatam says that it is our attachment to sense objects of the physical plane that tricks us: "Because of identifying with the material activities that are in fact performed by the modes of nature, does the living entity thus wrongly attribute them to himself. From the misconception of its life in material conditioning is it made dependent..."[3]

The continued focus on sense objects creates samskaras that create more attachment and bind us to the physical world through karma and future rebirths. All the thoughts and actions applied to sense objects give strength to them and the material plane. We cannot have heaven upon earth if all we see is the material world and not the divinity within it.

Dispensationalists and many fundamentalists devoted to a God of Wrath put even more pressure on Mother Earth as they wait for her destruction as Barbara Rossing notes:

The world cannot be saved—that is the basic Rapture credo, proclaimed by televangelists, radio preachers, and best selling end times thrillers. Rapture proponents seem willing to live in the world with no more responsibility for caring for it than just letting the clock run out. They love to cite statistics about how the world is getting worse: crime is on the increase, wars and earthquakes are more frequent, the oceans are polluted, and environmental degradation is worsening. To them, these 'signs' prove that the prophetic clock has counted down almost all the way and then they can soon escape.[4]

The focus on escapism and destruction combined with praying for the end of the world creates a hideous thought form that casts a pall over the entire world. The signs that they see of Armageddon—increasing wars, earthquakes and environmental degradation—are reflections in the mirror of their own consciousness. They are part of the collective consciousness of abusing Mother Earth and justifiable war and violence. While they may seek the salvage of individual souls they often fail to see God in all and misuse Mother Earth. The scorched earth mentality of end timers amounts to Matricide and is a grave risk.

A loving, caring and nurturing alternative must be provided to the materialism and destructiveness in the minds of so many in the world today. It is critical to believe in and give strength to the kingdom of heaven upon earth. We need to dream of and pray for the Promised Land. In doing so we will be creating a loving and caring thought form that will on its own find a way to nurture and help those suffering in the material world.

Pray for Heaven Upon Earth

The prophets have told us that we will have heaven upon earth when violence ends, love abounds and when the lion lies down with the lamb. It will be a place where justice, compassion and mercy rein; a place where all in God's kingdom live together in peace and happiness.

Pray and meditate for heaven upon earth. Visualize an alternative world where peace, happiness, joy and love permeate the air and consciousness of all living beings. Incorporate the vision of heaven upon earth into your life by seeing heaven unfolding before your eyes. See the little miracles in your daily life of compassion, love and kindness as a sign that heaven is beginning to manifest on earth.

Meditate on and visualize God's light and love shining upon those that preach the destruction of Mother Earth and to those harming her with excavation equipment, chain saws and large construction equipment. Visualize warring parties laying down their weapons and reaching out to hug each other. See mother earth healing. Visualize happy glowing faces gathering to pray and to honor mother earth.

Think of a beautiful moment of love, happiness and compassion in your life and the emotion it brought and then attach it to your dream of heaven on earth. Well over with love and joy as you see heaven unfolding before you. Feel the joy and blessing of God for sharing this with you. Believe it. See it.

Let the Dream Move You

While you are meditating on heaven on earth you might be moved or even be called to action. Should you feel called to do something you should meditate on it; think, reflect, and contemplate on it. If it is meant to be God will keep bringing you the vision. If it comes up again take it to heart. Look for direction and inspiration as to how you are supposed to manifest your vision. Mediate and contemplate upon it more and visualize it becoming a reality. Thank God for the inspiration.

It was Martin Luther King's dream of seeing the constitution that "all men are created equal" become a reality that drove him and a movement that transformed America. It was a vision of God's word becoming a reality, "But let justice roll down like waters, and righteousness like an ever-flowing stream"[5] that inspired and drove him.

As we move forward in our dreams many will be called. Heed that word, it is the manifestation of our thoughts and prayers and God's divine will.

Attachment

It is our attachment to transient things whether they are sense objects, organizations, ideas or the illusion of an independent self that causes suffering and gives strength to the physical world. The First Noble Truth of Buddhism is that "Life is suffering," The second Noble Truth is that "the origin of suffering is attachment." Attachment binds us to the material world. 1 John tells us that if we love the material room there is no room left to love God: "Do not love the world or the things in the world. If anyone loves the world, the love of the Father is not in him."[6]

There are a variety of spiritual exercises that you can undertake to help you break free from the material world of sense objects. Meditation is a good beginning. I have been given and used mantras like, 'the world is not real' ('it's not real, it is an illusion') to help me overcome the belief that the physical world is real and understand that it is an illusion cloaked in maya. Repeating such a mantra will help you extricate yourself from the physical plane and realize that there is another true reality.

You can also meditate on nothingness (void) or outer space. Imagine what the world not cloaked in maya looks like. Or keep your eyes open and meditate on the sky. By practicing these meditations we give strength to an alternative and reduce the pull of the material world.

Renunciation

We must embody the spirit of non-attachment through the action of renunciation in our daily lives. We must endeavor to follow the advice of Jesus who told the rich man that he had to abandon all of his wealth and its trappings if he wanted to get to heaven.[7] If you dream of and believe in the possibility of another world, a heaven upon earth, then your actions should similarly reinforce this. Start by reducing the strength that you give to the material world by focusing less on the world of sense objects. Be selective about the things that you buy, the events you participate in and the places you go. Renunciation is about beginning a withdrawal process from the material world. We must, as we do with meditation, endeavor to shut out the physical world around us.

Renunciation, or non-attachment, or non-involvement should not be confused with being against something. Renunciation is not about protesting or being against anything, but rather is about the refusal to participate in something, to be un-involved, to not own, to not partake.

Renunciation has historically been associated with the monastic, or ascetic life. However that is not possible for most of us. Instead we should as William Penn suggests carry the cloister, the monastery or ashram in our heart and with our daily interactions. "The Christian convent and monastery are within,

225

where the soul is encloistered from sin; and this religious house the true followers of Christ carry about with them, who exempt not themselves from the conversation of the world, though they keep themselves from the evil of the world in their conversation."[8]

As noted earlier the Bhagavad-Gita says that we can detach ourselves from the material world by being indifferent to the fruits of our actions.[9] This involves participating in life more as an observer than a participant in the physical plane. By not attaching emotion to an object or an action we are not giving it the same power of thought as we would if we were emotionally involved. We have to learn how to be in this world, but not of it.

Walking the Path of Renunciation

We, individually and collectively, need to begin a process of renunciation in which we do not own certain sense objects, participate in certain events or associate with certain groups. This is the beginning of a longer term effort to make the life of the sangha, the ashram and the monastery, the mainstream.

The false gods of science and the market continually cast a web to catch our attention. Currently it is impossible to escape their clutches. We use technology to travel, communicate, heat our homes and cook our food. We are bombarded by the market and forced to use money for exchange. We participate in the market system through work, the purchase of basic necessities like food, through our communications and through our entertainment. We are trapped. It is near impossible to escape.

What is possible is to start walking the path of renunciation by slowly beginning to reduce the influence of false gods and malicious thought forms in your life. Selectively choose what you are going to give strength to through your thoughts and actions. View it as you would a diet to loose weight by changing your life style. Only here you are beginning to stop consuming the material world and the temptations of false idols. This will be a slow process that will move in baby steps. The material world has been able to grow almost unabated for thousands of years and we cannot extract ourselves immediately. Do not get frustrated with setbacks or challenges.

Look within yourself to determine how the process of renunciation is going to unfold in your life. Meditate on it. Ask God for help. Your best guide is within yourself. Visualize and develop a thought form to help you through the process. Reflect on it as you go forward to give it strength. There are no hard and fast rules about what you should, or should not, take out of your life. You must search your soul and decide where to begin and what to cut out. Some of you may want to stop eating meat immediately, others may get rid of their TV. Others may be unable to get rid of their fancy car, or eliminate watching sporting events.

I suggest you begin looking at how you are manipulated by the system, to see how you are programmed to consume or to think in a certain way. Resist buying the latest gadget or the newest fad in clothing. Avoid seeing the latest and hottest movie, or buying the most popular video or CD. Try not to give strength to our culture of hot, hip and cool, by thinking or talking about it. Instead focus on heaven on earth, God or something else spiritual.

Pay no attention to Hollywood shape shifters, or idols in other industries such as business, sports or media. Think about how you can begin to extricate yourself from popular culture. Remember that the losers in the system are the winners in the kingdom of heaven. So don't be a slave to fashion. Resist.

Activities and Objects

Remove violent things whether they be guns, violent movies, violent video games and violent sports from your life as soon as possible. Do not participate in the military, its weapons or air shows, ceremonies, or other entrapments. Remove objects associated with hate, prejudice, and sexism from your life.

Slowly begin to remove yourself from competitive sports. Try not to identify with your college or hometown team; do not spend hours at a game, or in front of your TV getting excited and/or angry. Try not to participate in competitive sports. Exercise your body but not in games where you are pitted against someone else.

Reflect on your identity. I am a first generation Estonian who did not learn to speak English until I was five years old and grew up in a very tight knit ethnic community. One day in the mid 1980's I was watching the TV news and saw all the bloodshed in the Middle East, South Africa and in Northern Ireland. At that moment I felt that my ethnic association was contributing to the ethnic, religious and racial violence in the world. While it was impossible for me to totally eradicate my Estonian roots it was possible for me to not attend Estonian events or not identify with my ethnic roots.

Do not give strength to your ethnic, racial or cultural background. Instead see yourself and others as being part of God. Understand that who you are in this life could be something different in the next life. We are all reincarnated souls, the white person was and will become a black person some day and vice versa, the Muslim was and will be the Jew and vice versa.

We all need to look at the religious and other institutions we belong to. Participation in such organizations gives strength to artificial constructs that take on a life of their own, independent of and often in conflict with the principles espoused by their founder, or prophet. Do not participate. Have faith that something better will come into your life.

It is important not to confuse tools with toys, or make tools toys. All technology is meant to be a tool to improve our lives. Unfortunately some of us have turned tools into toys that we worship and live for. For some people a car, a motorcycle, a boat, an ATV, a computer have become pleasure objects by which they often define themselves by. You may think that your car is a reflection of your taste in style, but in reality you are an appendage of your car. Such behavior is a blatant manifestation of how the false gods of science and technology draws us in.

Reduce technology in your life. Don't have every latest gizmo ever created. Get rid of the talismans in your home that are representative of an ostentatious or extravagant lifestyle. Keep to basic necessities and avoid objects or wealth, power and our system of domination.

A multi-volume book could be written on the process of renunciation providing great detail about the choices, dilemmas and challenges. I have only provided a very brief overview. You need to look within yourself to determine how the process is going to unfold in your life. There are certain things that you can let go of quickly, others that you may never let go of. Look within yourself. You are facing at a lifelong process.

Renunciation is a slow and gradual process to remove and detach ourselves from the physical plane of sense objects. Have faith that void left in your life will be filled with something wonderful. New associations will develop and come and go, as will new ways of practicing your faith, as we move forward on our sojourn. As they say you have to get rid of the old to have room to let the new in. The process of renunciation will ultimately herald dramatic change in the physical world.

Take inspiration from the Jains. One of their central tenets is Aparigraha, the non-attachment to objects and people, it espouses that one should only have the bare necessities in life. It is said that Mahavira who gave Jainism many of its central tenets had no possessions at all, not even clothes to cover his body.

Jesus teaches us that when we do not participate in the material world we will be scorned; "If the world hates you, you know that it hated Me before it hated you. If you were of the world, the world would love its own. Yet because you are not of the world, but I chose you out of the world, therefore the world hates you."[10]

Putting forth the Spiritual, the Positive

Renunciation should be accompanied with putting forth the spiritual and the positive into the world. We have already discussed this with our thoughts and actions in bringing heaven on earth. We need to supplement that.

We need to provide an alternative to the false gods of science and the market and all the other hideous thought forms that are bereft of any higher values and bring the human race to the depths. This needs particularly to be done in our schools. People talk about the separation of church and state not realizing that science and capitalism are religions. Children should not be taught one faith is supreme, but about all teachings. They should learn how to meditate and other spiritual exercises such as prayer. Meditation can help students to relax and concentrate better; it will also help raise their consciousness. Meditation will help them understand their thoughts. It will help them tap into Mother Earth and in doing so raise their consciousness and become healthier. Meditation can be a powerful tool to help transform the world and bring about heaven on earth.

An alternative to the competition of sports needs to be developed. We have to shift the focus of physical activity and exercise to health and spiritual benefits and not be a breeding grounds for competition.

Government needs to be refocused back on community and working together rather than satisfying special interest. Our emphasis needs to be on inner exploration and not space exploration and or scientific investigation to better exploit the environment. Our spending and focus needs to be on the heart and helping and not on violence, military and conquest. Special interests need to be abolished whether they be corporations, unions, religions or NGO's. We have to empower people and not organizations.

The positive alternatives are not about rechanneling the market or science, but rather a new way of doing things. It is about putting forth God, spirituality and making heaven upon earth. New amalgams, teachers, movements and directions will spring up. We just need to have faith that we will be delivered and believe that something new and better will be coming.

Imitation, the example of Jesus

The prophets serve as an example of renunciation and right living. The Muslim Hadith tells the story of Mohammed's life and how Muslims are to live. The Bible says that we are to imitate Christ.[11] The desert hermits were early followers of Jesus who renounced the material world for the desert. Thomas a Kempis wrote *The Imitation of Christ* that gave us a set of spiritual exercises we are to follow to be Christ like.

The story of Jesus personifies God's call to love, ahimsa (non-violence towards all) and the renunciation of the material world and all its trappings. While those of faith rightly look to Jesus for inspiration, he was the antithesis of the material world. Jesus was a tragic hero, arguably a failure in the world of materialism and greed.

Jesus constantly advocated love and told us that we should even love our enemy[12] and turn the other cheek if we are struck.[13] He taught that we are too give love in response to violence. He embodied his words through action. When the Roman soldiers came to arrest him and Peter resisted and cut off a soldiers' ear, Jesus told him to stop and healed the solider. Jesus voluntarily let them take him away. He did not resist violence and died on the cross asking forgiveness for those that had injured him: "Father, forgive them; for they do not know what they are doing."[14]

He defied convention and the material world. He told the rich man that to get to heaven he should abandon all his wealth and possessions and follow him.[15] He showed that the pursuit and possession of wealth to be an apostasy when he said: "It is easier for a camel to go through the eye of a needle than for someone who is rich to enter the kingdom of God."[16]

Jesus was not the stereotypical hero portrayed in novels and movies. There was no marching band, no extravagant entourage accompanying him when he entered Jerusalem. He walked in humbly and was carried by a donkey. He was mocked and jeered as the king of the Jews and given a crown of thorns to wear before his crucifixion.

In a world of materialism and violence Jesus was clearly a 'loser' in that world. Through example he showed us that we should abhor violence and embrace love in all and in everything. His actions demonstrate that we should similarly renounce the material world and possessions. He is a shining of how we should live.

Jesus through his submission to the Roman soldiers that came to arrest him and crucifixion shows that violence, even for resistance, is never justified. War, aggressive acts for national defense or violent acts in the name of God such as suicide or terrorist bombings are wrong and defile God and Jesus. Soldiers, suicide bombers and all those that employ violence and kill are not only defiling God but attacking God and themselves. Jesus consistently taught nonviolence[17] and advocated transforming those that persecuted you and were against you by turning the cheek through non-violent resistance.

Chapter 14

Conclusion- Our Faith Can Deliver Us

The opportunity for a better world of peace and happiness, a heaven on earth, is within our grasp. To do this we need to shift our consciousness from the physical plane of existence and the trappings of the material world to higher planes of existence and the divine. It is time we lived the life of the spiritual beings that we are.

What I am proposing is very radical, arguably unbelievable to some, but it is what we are called to do if we are to liberate ourselves. The challenge is enormous. We have been wedded to the world of sense objects for so long that it has clouded our thinking and produced numerous addictions for ourselves. We have created false gods like science and the market that manipulate and deceive us. All of this will pull at us as we try to extricate ourselves.

By planting thought seeds of love and divinity in the world we will begin our walk to the Promised Land like the Israelites freed from chains of slavery in Egypt. We must endeavor to anchor our hearts and minds in heaven as our bodies traverse the material world. It is a journey that we must take collectively, one step at a time. We are also beginning a journey without everyone aboard and there are many out there who will be casting doubt. We have to believe and pray that others will eventually join the journey.

I do not know how this process is going to unfold or how things will change. I do know that it will take a very, very long time. No doubt mysterious things can and will happen. The greatest hope for change comes from rekindling our relationship with Mother Earth. She is so ravaged that the smallest effort can bring about the greatest changes. It is also imperative that we focus on her because her health teeters from of our violent thinking and behavior. We should be prepared for the physical world to dramatically change as the mystical begins to manifest. What we call reality will be greatly transformed.

It is our faith that will deliver us. Understand that there is no visible path before us, only the belief that there is a path. Have confidence that our faith will persevere and guide us. Trust and develop your faith, sprinkle it with the yeast of commitment and dedication so that it may rise higher and higher.

As we move forward new challenges will come. Spiritual insights, perceptions and abilities will develop as reality begins to filter through the illusion. This blessing will bring about new temptations and new pitfalls. We must be like Ulysses looking to return home to his beloved Penelope and not be trapped by what we encounter on the way home.

Our greatest hope is that what appears miraculous and virtually an impossible path today, future generations will look back and say how backwards and simple our vision was. Then it will be evident that the seed thoughts we planted today will have born the most exquisite fruit. Let us pray that this becomes so. Have faith that if we live righteously and love each other and God's creatures and Mother Earth, that we will be delivered to the Promised Land of milk and honey that Hosea spoke of,

Therefore, I will now persuade her, and bring her into the wilderness, and speak tenderly to her. From there I will give her vineyards, and make the Valley of Achor a door of hope…

I will make for you a covenant on that day with the wild animals, the birds of the air, and the creeping things of the ground; and I will abolish the bow, the sword, and war from the land; and I will make you lie down in safety. And I will take you for my wife for ever; I will take you for my wife in righteousness and in justice, in steadfast love, and in mercy.

I will take you for my wife in faithfulness; and you shall know the LORD.[1]

That Promised Land awaits us. The place where the lion lies down with the lamb, where we are never of want and peace is the way. Believe in it and remember that we live in a mystical world.

Notes

Introduction

1. Buddhist term for liberation.
2. Hindu term for liberation from the cycle of life and death.
3. Genesis 2.10-14.
4. Exodus 3.8.
5. Isaiah 11.6, 9.
6. Matthew 6.10.
7. Matthew 6.14-34.
8. Matthew 5.5.

Chapter 1 Hidden World(s)

1. Sri Aurobindo (Ghose), *Collected Works Volume 20, Synthesis of Yoga*, Sri Aurobindo Ashram Trust, http://sriaurobindoashram.info/Default.aspx?1=234, 1970; 69.
2. I would describe our sixth sense as all of our other senses beyond our five physical senses, the ability to experience God, intuition, clairvoyance, ability to see aura's and the like.
3. Thomas Merton, *The Inner Experience, Notes on Contemplation*, HarperSanFrancisco, 2003; 35.
4. Svetasvatara Upanishad 4.10.
5. The defining and understanding the concept of maya is illusive and it opens up a Pandora's Box of sorts. If we believe all is an illusion and detachment is the remedy then we can slip into the misconception that nothing exists and everything is a figment of our imagination. This can lead to extreme introversion and withdrawal and ultimately undermining the reality that all is Brahman and our unity and oneness; in doing so we ignore the message of the Hebrew prophets who taught social justice and care for the poor.
6. Dhammapada 170.
7. Michael Talbot, *The Holographic Universe*, Harper Perennial, New York, NY 1992; 1.

8. William J. Cromie, "Meditation Changes Temperature: Mind Controls Body in Extreme Experiments" Harvard Gazette, April 18, 2002.

9. Luke 17.20,21.

10. "The total material substance, called Brahman, is the source of birth, and it is that Brahman that I impregnate, making possible the births of all living beings, O son of Bharata." Bhagavad-Gita 14.3
"I am the father of this universe, the mother, the support, and the grandsire. I am the object of knowledge, the purifier and the syllable om. I am also the Rig, the Sama, and the Yajur [Vedas]". Bhagavad-Gita 9.17.

11. Chandogya Upanishad VI.9.1-4.

12. The Essene Gospel of Peace, Book One; Page 9.

13. Fritjof Capra, The Tao of Physics — An Exploration of the Parallels between Modern Physics and Eastern Mysticism, Shambala, Boston, 2000; 131.

14. Chandogya Upanishad 6.2.3.

15. My meditations within Mother Earth's soul have led me to believe that there is something beyond consciousness associated with higher planes of existence. Within Mother Earth's soul there are fields of consciousness that emanate consciousness, but there are other things there as well. It appears that these attributes of higher planes manifest, or become discernable, in areas were extensive spiritual practices have taken place.

16. Mundaka Upanishad II.1.8.

17. In his commentary of Vishnu Purana I.6 Horace Hyman (48.10) describes the various lokas. Hindu Vishnu Purana, translated by Horace Hayman Wilson, 1840; http://www.sacred-texts.com/hin/vp/index.htm.

18. Swami Satyasangananda Saraswati, Sri Vijnana Bhairava Tantra, The Ascent, Yoga Publications Trust, Munger, Bihar, India, 2003; 19-20.

19. Bhante Suvvanno, *The Thirty One Planes of Existence*, Jinavamsa [Inward Path] Publishing, Penang Malaysia, 2001, http://www.buddhanet.net/pdf_file/allexistence.pdf.
20. H.P. Blavatsky, *Collected Works, Volume 12,*: EI: Instructions No. IV. 656-673.
21. Swami Vivekananda, *Raja Yoga*, Ramakrishna-Vivekananda Center, New York; 76.
22. The second of the Four Noble Truths.
23. Bhagavad-Gita 2.28.
24. H. P.Blavatsky, *The Key to Theosophy—H. P. Blavatsky and Abridgement*, Edited by Joy Mills, Quest Publishing, Third Edition 1992; 121-122.
25. H. P.Blavatsky, *The Key to Theosophy—H. P. Blavatsky and Abridgement*, Edited by Joy Mills, Quest Publishing, Third Edition 1992; 123.
26. Hosea 1.4. In 2 Kings 9-10 tells how Jehu, supposedly under God's auspices had Ahab, his descendant's and court wiped out.
27. Rose G. Lurie, *THE GREAT MARCH Post-Biblical Jewish Stories BOOK I*, Illustrations by Todros Geller, The Union of American Hebrew Congregations, YORK, 1931, http://www.sacred-texts.com/jud/tgm/tgm00.htm.
28. John 3.3.
29. "For the soul there is never birth nor death. Nor, having once been, does he ever cease to be. He is unborn, eternal, ever-existing, undying and primeval. He is not slain when the body is slain." Bhagavad-Gita 2:20.
30. Swami Vivekananda, *What Religion Is*, The Julian Press, New York, 1962; 74.
31. Swami Niranjanananda Saraswati, *Prana Pranayama, Prana Vidya*, Bihar School of Yoga, Munger, Bihar, 1998; 4.
32. Theosophists see the etheric plane being closer to a higher sub plane of the physical plane.
33. Madame Blavatsky the founder of theosophy calls the astral

plane the karmic plane.

34. C.W. Leadbeater, *The Astral Plane*, The Theosophical Publishing House, 1973; 1-2.

35. Sri Aurobindo (Ghose), *Collected Works Volume 24, Letters on Yoga Part 4*, Sri Aurobindo Ashram Trust, http://sriaurobindoashram.info/Default.aspx?1=234, 1970; 1368.

36. Arthur E. Powell, *The Mental Plane; The Mental Body*, First published in 1927 by The Theosophical Society, www.anandgholap.net/Inner_Life_Vol_II-CWL.htm; Chapter XVIII.

37. Consciousness might not be the right word to describe our evolution because it may well be something beyond consciousness.

38. Theosophists call the next plane of existence above the mental plane the Buddhic plane, or the plane of spiritual love. Sri Aurobindo calls the level above the mental plane as the Supermind, or infinite truth consciousness. According to Theosophists the atmic plane, or spiritual will resides above the buddhic plane. C. W. Leadbeater calls it the nirvanic plane and says that most of us are incapable of understanding it (*The Devachnic Plane*; 102.) Theosophists believe that the monadic plane, the monad or higher self is found above the Atmic and below the highest plane, the adi Plane or divine plane.

39. Sri Aurobindo calls it the vital. Theosophists see it consisting of two parts the etheric double (which deals with energy) and the astral body (emotion and desire).

40. Swami Satyasangananda Saraswati, *Sri Vijnana Bhairava Tantra, The Ascent*, Yoga Publications Trust, Munger, Bihar, India, 2003; 149.

41. C. W. Leadbeater, *Inner Life Volume 1*, Theosophical Publishing House, Adyar, India, 1917; 638.

Chapter 2— Thoughts, Thought Forms, Samskaras

1. Siri Guru Granth Sahib, page 6.
2. Bhagavad-Gita 6:6.
3. Bhagavad-Gita. Chapter 16, "The Divine And Demoniac Natures."
4. William Walker Atkinson, *Thought Vibration or the Law of Attraction in The Thought World*, First Published by The New Thought Publishing Co. Chicago 1906, Electronic Edition Published by Cornerstone Publishing, 2001; Chapter 1.
5. Frances Vaughan, *Shadows of the Sacred; Seeing Through Spiritual Illusions*, Quest Books, The Theosophical Publishing House, Wheaton, Illinois, USA, 1995; 255.
6. Frances Vaughan, *Shadows of the Sacred; Seeing Through Spiritual Illusions*, Quest Books, The Theosophical Publishing House, Wheaton, Illinois, USA, 1995; 255.
7. Gustave Le Bon, *The Crowd, A Study of the Popular Mind*, T. Fisher Unwin, London, 1903; 143.
8. Annie Besant and C. W. Leadbeater, *Thoughts Forms*, Theosophical Publishing House, Wheaton, Ill., 1971, c1925; 18.
9. Annie Besant and C. W. Leadbeater, *Thoughts Forms*, Theosophical Publishing House, Wheaton, Ill., 1971, c1925; 22.
10. I must admit that for a few years in the 1990's I hunted birds. For several years afterwards I mistakenly believed that it was more humane for a hunter to take a deer than have it suffering a slow death over the winter or at the hands of a pack of coyotes. As I have come to learn about how our thoughts work in the universe it is very clear that hunting encourages predatory behavior. Hunters must be told and educated about the ramifications of their behavior.
11. Matthew 5.27-28.
12. James Allen, *As a Man Thinketh* contained within *The Wisdom of James Allen, 5 Classic Works Combined in one*; Radiant

Summit Books, San Diego, California, 1997; 23.

13. Steady gazing at an object; meditating with your eyes open.

14. The circular pattern of consciousness.

15. Swami Satyananda Saraswati, *Four Chapters on Freedom—Commentary on the Yoga Sutras of Patanjali*, Yoga Publications Trust, Munger, Bihar, India, 2002; 111.

16. Swami Satyasangananda Saraswati, *Sri Vijnana Bhairava Tantra, The Ascent*, Yoga Publications Trust, Munger, Bihar, India, 2003; 292.

17. Frances Vaughan, *Shadows of the Sacred; Seeing Through spiritual Illusions*, Quest Books, The Theosophical Publishing House, Wheaton, Illinois, USA, 1995; 256.

18. Roger Highfield, "Mankind 'Shortening the universes life'", UK Telegraph, November, 21, 2007.

19. Lama Thubten Yeshe, *Introduction to Tantra, A Vision of Totality*, Wisdom Publications, Boston, Ma, 1987.

20. Thich Nhat Hanh, *Creating True Peace*, Free Press, New York, 2003; 11.

21. H.P. Blavatsky, *Isis Unveiled, A Master-Key to the Mysteries of Ancient and Modern Science and Theology Volume 1*, Theosophical University Press, Pasadena Calif.,1877, http://www.theosociety.org/pasadena/isis/iu-hp.htm; 244.

22. William Walker Atkinson, *Thought Vibration or the Law of Attraction in The Thought World*, The New thought Publishing Company, Chicago, 1906; 1-2.

23. Mirra Alfassa, The Mother, *The Mother; Collected Works, Volume 9*, Sri Aurobindo Ashram Trust, http://sriaurobindoashram.info/Default.aspx?1=234, 1972; 386.

24. There are other methods and practices that can clean up samskaras such as prayer and meditation as well as God's grace.

25. Swami Satyananda Saraswati, *Four Chapters on Freedom—Commentary on the Yoga Sutras of Patanjali*, Yoga Publications Trust, Munger, Bihar, India, 2002; 139

26. Swami Vivekananda, *What Religion Is,* The Julian Press, New York, 1962; 51.

27. "[S]amskaras are the source of bondage and keep man forever chained to the cycle of birth and death." Swami Satyasangananda Saraswati, *Sri Vijnana Bhairava Tantra, The Ascent,* Yoga Publications Trust, Munger, Bihar, India, 2003; 281.

28. Swami Vivekananda, *What Religion Is,* The Julian Press, New York, 1962; 53.

29. Mirra Alfassa, The Mother, *The Mother; Collected Works, Volume 6,* Sri Aurobindo Ashram Trust, http://sriaurobindoashram.info/Default.aspx?1=234, 1972; 279.

30. If you ever find yourself in a cycle of continual calamity you can attempt to counter it by giving. Take it upon yourself to help everyone and everything—help your neighbor or a stranger, work at a soup kitchen, volunteer at a hospital, visit someone in prison. Eventually the act and thought of giving will come back to you and help you overcome your downward spiral.

31. C.G Jung, *Collected Works Volume 10, Civilization in Transition, Volume 10,* Bollingen Series, Princeton University Press; 1968, c1958; "Man and Earth", 49.

32. Swami Satyasangananda Saraswati, *Sri Vijnana Bhairava Tantra, The Ascent,* Yoga Publications Trust, Munger, Bihar, India, 2003; 32.

33. Nick Ferrell, *Making Talismans, Living Entities of Power,* Llewellyn Publications, St. Paul Minnesota, 2001; 1

34. Isaiah 2.8.

35. Matthew 6.21.

36. "[O]ne of the best-kept secrets in medical science is the extensive experimental evidence for "spiritual healing." Daniel J. Benore, M.D., an American psychiatrist working in England, survey all such healing studies published in the English language prior to 1990. He defined "spiritual

healing" as "the intentional influence of one or more people upon another living system without utilizing known physical means of intervention." His search turned up 131 studies, most of them in no-humans. In fifty-six of these studies, there was less than one chance in hundred that the positive results were due to chance. In an additional twenty-one studies, the possibility of a chance explanation was between two and five chances in a hundred" Larry Dossey, M.D., *Healing Words, the power of Prayer and the Practice of Medicine*, Harper San Francisco, 1993; 189

Chapter 3 Sweet Mother Earth

1. Sri Aurobindo (Ghose), *Collected Works Volume 22, Letters on Yoga Part 1*, Sri Aurobindo Ashram Trust, http://sriaurobindoashram.info/Default.aspx?1=234, 1970; Page 17.
2. Geoffrey Hodson, *The Kingdom of the Gods*, The Theosophical Publishing House, Adyar, India, 1981; 61.
3. James Lovelock, *Healing Gaia, Practical Medicine for the Planet*, Harmony Books, New York, 1991; 153.
4. Jiddu Krishnamurti, *All the Marvelous Earth*, Krishnamurti Foundation of America, Ojai, California, 2000; 36.
5. Jane Thurnell-Read, *Geopathic Stress—How Earth Energies Affect Our Lives*, Element Books, Rockport, Ma, 1995; 31.
6. Because of the earth's rotation there are differences in the direction of spin for positive movement in the northern hemisphere from the southern hemisphere.
7. See http://www.gaiassoul.org/.
8. The history of the thoughts and actions around a Line of Consciousness will determine the influence beyond the line.
9. http://www.motherearthprayers.org/.
10. http://www.jubileeinitiative.org/SacredPropheticSpirit.html.
11. Swami Niranjanananda Saraswati, "Yoga and Total Health", Yoga Magazine (Bihar School), January 2002; http://www.yogamag.net/archives/2002/1jan02/totheal.shtml

12. Genesis 6.5-8.
13. Jeremiah 3.9.
14. "The wicked are not so, but are like chaff that the wind drives away.", Psalm 1:4.
15. Jeremiah 23.10.
16. Swami Niranjanananda Saraswati, *Prana, Pranayama, Prana Vidya*, Bihar School of Yoga; Munger, Bihar; 1998; 86.
17. Swami Vivekananda, *Raja Yoga*, Ramakrishna-Vivekananda Center, New York, 1973; 102.
18. Sun Bear and Wabun Wind, *Black Dawn Bright Day*, A Fireside Book, New York, 1992; 29.
19. Kundalini energy is the highest spiritual energy of the divine. It facilitates our spiritual evolution.
20. Shyman Sundar Goswami, *Layayoga—The Definitive Guide to the Chakras and Kundalini*, Inner Traditions, Rochester, Vermont, 1999; 84.
21. Swami Satyadharma, *Yoga Chudamani Upanishad, Crown Jewel of Yoga*, Yoga Publications Trust, Munger, Bihar, India, 2003; 115.
22. "There have been other dowsers who have also noticed the apparent ability of new labyrinths that were used regularly, to draw water to them - John Wayne Blassingame and Marty Cain, both good dowsers in the US - to mention two. Here at Benton, was the first time I had seen that energy leys were also drawn to the newly constructed space." Sig Lonegren, "The Gathering Earth Energies: An Interim Report", Mid-Atlantic Geomancy; Winter Issue 1996, http://geomancy.org/e-zine/1996/winter/gathering-earth-energies/index.html.
23. Steven Skinner, *The Living Earth Manual of Feng Shui*, Routledge & Kegan Paul Ltd., London, 1982; 4.
24. Sun Bear and Wabun Wind, *Black Dawn Bright Day*, A Fireside Book, New York, 1992; 172.

Chapter 4 Inhabitants of Other Worlds

1. *The Sutra of Hui Neng.*
   http://www.sinc.sunysb.edu/Clubs/buddhism/huineng/
   content.html; Chapter X: "His Final Instructions."

2. Mara is the demon who tempted Buddha by trying to seduce
   him with visions of women.

3. *The Sutra of Hui Neng,*
   http://www.sinc.sunysb.edu/Clubs/buddhism/huineng/
   content.html; Chapter X: "His Final Instructions."

4. Alexandra David-Neel, *With Mystics and Magicians in Tibet,*
   John Lane The Bodley Head, Ltd, London, England; 1931;
   299-300.

5. Mirra Alfassa, The Mother, *The Mother; Collected Works,
   Volume 5,* Sri Aurobindo Ashram Trust, http://sriau-
   robindoashram.info/Default.aspx?1=234, 1972; 155.

6. Annie Besant and C. W. Leadbeater, *Thoughts Forms,*
   Theosophical Publishing House, Wheaton, Ill., 1971, c1925;
   15-16.

7. Alexandra David-Neel, *With Mystics and Magicians in Tibet,*
   John Lane The Bodley Head, Ltd, London, England, 1931;
   308-313.

8. Aryeh Kaplan, *Sefer Yetzirah—The Book of Creation, In Theory
   and Practice,* Weiser Books, Boston, Ma. 1997; 127, 128,130.

9. C. W. Leadbeater, *The Astral Plane,* The Theosophical
   Publishing House, 1973; 138-139.

10. Theosophists hold that our body fragments upon death into
    our visible body, our emotional body and our higher self. The
    visible and emotional bodies began to decay almost immedi-
    ately. See H. P. Blavatsky, *The Key to Theosophy—H. P.
    Blavatsky and Abridgement,* Edited by Joy Mills; Quest
    Publishing, Third Edition 1992; Chapter IX "The Kama Loka
    and Devachan."

11. What he would refer to as the Astral Plane, or the Astral
    component of the Pranic Plane.

12. C. W. Leadbeater, *The Astral Plane*, The Theosophical Publishing House, 1973; 6-7.

13. Sri Aurobindo (Ghose), *Collected Works Volume 22, Letters on Yoga Part 1*, Sri Aurobindo Ashram Trust, http://sriaurobindoashram.info/Default.aspx?1=234, 1970; 395.

14. It takes time, effort and hard work hard to have protection work.

15. On several occasions I have encountered visitors when doing energy work (group meditation/pranayma) at some place other than a dedicated place of worship. Rarely have I encountered visitors doing the same at a place of worship. This is not to say that doing energy work at a place of worship dedicated to God is fool proof. It helps.

16. Harriet Maxwell Converse, *Myths and Legends of the New York Iroquois*, edited and annotated by Arthur Caswell Parker, Ira J. Friedman, Inc., Port Washington, NY, 1962; 101.

17. Luke 1.26-38.

18. Exodus 3.1-2.

19. Exodus 23.20-23.

20. *The Kingdom of the Gods*, The Theosophical Publishing House, Adyar, India, 1980; 56.

21. Geoffrey Hodson, *Clairvoyant Investigations*, Quest, The Theosophical Publishing house, Wheaton, Illinois, USA, 1984; 8-9.

22. Chapter 4, Verse 7-8; I used an interpretation by Jayaram from Hinduwebsite.com because I thought it made clear that part of God's essence descends to earth. http://www.hinduwebsite.com/

23. The Essene Gospel of Peace, Book One; 46.

Chapter 5 Seed Thoughts and Their Fruit

1. Bhagavad-Gita 2.62, 63.

2. Bhagavad-Gita 2.60.

3. A.C. Bhaktivedanta Swami Prabhupada, *Bhagavad-Gita(As it*

is); The Bhaktivedanta Book Trust, Los Angeles, 1996; commentary to 2:60, 79.

4. Charles Darwin, *The Origins of Species by Means of Natural Selection—Or the Preservation of Favored Races in the Struggle for Life,* from *The Origins of Species by Means of Natural Selection and The Descent of Man and Selection in Relation to Sex,* The Modern Library, New York; 52.

5. Charles Darwin, *The Origins of Species by Means of Natural Selection—Or the Preservation of Favored Races in the Struggle for Life,* from *The Origins of Species by Means of Natural Selection and The Descent of Man and Selection in Relation to Sex,* The Modern Library, New York; 64.

6. Charles Darwin, *The Descent of Man and selection in relation to sex,* from *The Origins of Species by Means of Natural Selection and The Descent of Man and Selection in Relation to Sex,* The Modern Library, New York; 430-431.

7. Charles Darwin, *The Descent of Man and selection in relation to sex,* from *The Origins of Species by Means of Natural Selection and The Descent of Man and Selection in Relation to Sex,* The Modern Library, New York; 431.

8. Walter Wink, *Engaging the Power, Discernment and Resistance in a World of Domination,* Fortress Press, Minneapolis, MN, 1992.

9. Walter Wink, *Engaging the Powers, Discernment and Resistance in a World of Domination,* Fortress Press, Minneapolis, MN, 1992; 33.

10. Dave Anderson, "Breaching the Field of Play, Sports Tumbled Out of Bounds," NY Times, December 30, 2007.

11. Michael S. Schmidt and Duff Wilson, "Report Ties Star Players to Baseball's 'Steroids Era," NY Times, December 14, 2007.

12. Judy Battista, "Sideline Spying: N.F.L. Punishes Patriots' Taping," NY Times, September 14, 2007.

13. Thane Rosenbaum, *The Myth of Moral Justice, Why our Legal*

System Fails to do What's Right, Harper Collins, New York, 2004; 17-18.

14. Matthew 22.36-40.

15. Donald Grinde in *The Iroquois and the Founding of the American Nation, Exemplar of Liberty: Native America and the Evolution of Democracy* and Bruce Johansen *Forgotten Fathers; How the American Indian Helped Shape Democracy* have documented the influence of the Peacemaker, the Great Law of Peace and the Haudenosaunee in shaping American democracy.

16. The Great Law of Peace (Gayanashagowa) no.26.

17. "Axemakers, whose gifts change the world, have been running similar experiments on human society for as long as they have been doing unnatural acts like building shelters and cultivating the countryside. The result is that they have made modern Western perceptions different from others.", James Burke and Robert Ornstein, *The Axemaker's Gift—A Double Edged History of Human Culture,* G. P. Putnam's Son, New York, 1995;15.

18. James Burke and Robert Ornstein, *The Axemaker's Gift—A Double Edged History of Human Culture,* G. P. Putnam's Son, New York, 1995; 123-129.

19. James Burke and Robert Ornstein, *The Axemaker's Gift—A Double Edged History of Human Culture,* G. P. Putnam's Son, New York, 1995; 201-218.

20. David Suzuki, *The Sacred Balance—Rediscovering Our Place in Nature,* Prometheus Books, Amherst, New York; 1998; 191-192.

21. Huston Smith, *Why Religion Matters,* HarperSanFrancisco, 2001; 81.

22. Dr. Jon Kabat-Zinn, *Coming to Our Senses, Healing Ourselves and the World Through Mindfulness,* Hyperion, New York, 2005; 146-147.

23. Dr. Jon Kabat-Zinn, *Coming to Our Senses, Healing Ourselves and the World Through Mindfulness,* Hyperion, New York, 2005; 149.

Chapter 6 Collective Thoughts and False Gods

1. Irving Janis, *Victims of Group Think—A Psychological Study of Foreign Policy Decisions and Fiascoes*, Houghton Mifflin & Co., Boston, Mass., 1972; 3.

2. Irving Janis, *Victims of Group Think—A Psychological Study of Foreign Policy Decisions and Fiascoes*, Houghton Mifflin & Co., Boston, Mass., 1972; 9.

3. Irving Janis, *Victims of Group Think—A Psychological Study of Foreign Policy Decisions and Fiascoes*, Houghton Mifflin & Co., Boston, Mass., 1972; 5.

4. Alan Greenspan, "The Challenge of Central Banking in a Democratic Society," Given at the Annual Dinner and Francis Boyer Lecture of The American Enterprise Institute for Public Policy Research, Washington, D.C., December 5, 1996. http://www.federalreserve.gov/boarddocs/speeches/1996/199 61205.htm

5. http://www1.worldbank.org/economicpolicy/managing% 20volatility/contagion/index.html

6. Gustave Le Bon, *The Crowd: A study of the Popular Mind*, Dover Publications, Mineola, NY, 2002 (first 1895, English 1896); 4.

7. Gustave Le Bon, *The Crowd: A study of the Popular Mind*, Dover Publications, Mineola, NY; 2002 (first 1895, English 1896); 9.

8. Gustave Le Bon, *The Crowd: A study of the Popular Mind*, Dover Publications, Mineola, NY; 2002 (first 1895, English 1896); 111.

9. Gustave Le Bon, *The Crowd: A study of the Popular Mind*, Dover Publications, Mineola, NY; 2002 (first 1895, English 1896); 14.

10. Jennifer Van Bergen says that "President Bush has declared war...on the Republic of America." *The Twilight of Democracy, The Bush Plan for America*, Common Courage Press, Monroe, Maine, 2005; 5.

11. Gustave Le Bon, *The Crowd: A study of the Popular Mind*, Dover Publications, Mineola, NY; 2002 (first 1895, English 1896); 22.
12. Allen Guttmann, "Sporting Crowds" in *Crowds*, Edited by Jeffrey T. Schnapp & Matthew Tiews, Stanford University Press, 2006; 131-132.
13. Gustave Le Bon, *The Crowd: A study of the Popular Mind*, Dover Publications, Mineola, NY; 2002 (first 1895, English 1896); 38-39.
14. Gustave Le Bon, *The Crowd: A study of the Popular Mind*, Dover Publications, Mineola, NY; 2002 (first 1895, English 1896); 40.
15. Tim Carver, "Bush puts God on his side," April 6, 2003, BBC. http://news.bbc.co.uk/2/hi/americas/2921345.stm.
16. C. W. Leadbeater, *The Astral Plane*, The Theosophical Publishing House, 1973; 139-140.
17. "If the eighteenth century myth of origins ultimately destroyed the ancient gods, pagan and Christian, les progress became the new deities of the age, and the late eighteenth century definition of their attributes is in many ways canonical for modern times." Frank E Manuel, *The Eighteenth Century Confronts its Gods*, Harvard University Press, Cambridge Mass 1959; 11.
18. David Hawkin, *The Twenty First Century Confronts its Gods — Globalization, Technology and War*, State University of New York Press, Albany, 2004;4.
19. Christopher Catherwood, *Why Nations Rage, Killing in the Name of God*, Rowman & Littlefield Publishers, Lanham Maryland 2002;3.
20. Moses Maimonides, *The Guide of the Perplexed*, translated from the original Arabic text by M. Friedlander, George Routledge & Sons Ltd, New York, 1904; 50.
21. Isaiah 54.5.
22. Hosea 1.2.
23. Jeremiah 3.20.
24. Moshe Halbertal and Avishai Margalit, *Idolatry*, Harvard

University Press, Cambridge Mass, 1992; 11.

25. Jeremiah 2.13.

26. Moses Maimonides, *The Guide of the Perplexed*, translated from the original Arabic text by M. Friedlander, George Routledge & Sons Ltd, New York, 1904; 51.

27. Moses Maimonides, *The Guide of the Perplexed*, translated from the original Arabic text by M. Friedlander, George Routledge & Sons Ltd, New York, 1904; 51-5.

28. "Be sure of this, that no fornicator or impure person, or one who is greedy (that is, an idolater), has any inheritance in the kingdom of Christ and of God." Ephesians 5.5.

29. Exodus 20.5.

30. Amos 5.21-24.

31. Mark 3.31-35

32. Krishnamurti Foundation of America, http://www.kfa.org/.

33. Krishnamurti Foundation of America, http://www.kfa.org/.

34. Carl Jung, *Collected Works, The Archetypes and the Collective Unconscious, Volume 9*, Bollingen Series, Princeton University Press, 1971, c1959; 3-4.

35. Carl Jung, *Collected Works, The Archetypes and the Collective Unconscious, Volume 9*, Bollingen Series, Princeton University Press, 1971, c1959; 4-5.

Chapter 7 The Market (mob) as God

1. Harvey Cox, "The Market as God," Atlantic Monthly, March 1999.

2. Harvey Cox, "The Market as God," Atlantic Monthly, March 1999.

3. Bradford Long, "The Shape of Twentieth Century Economic History," NBER paper 7569, http://www.j-bradford-delong.net/tceh/2000/TCEH_1.PDF.

4. Karl Polanyi, *The Great Transformation*, Rhinehart & Co, 1944; 75.

5. David Loy, in the forward to his "Religion and the Market",

http://www.religiousconsultation.org/loy.htm.

6. Adam Smith, *The Wealth of Nations*, Random House, Modern Library Edition, 1937; 14.

7. Adam Smith, *The Wealth of Nations*, Random House, Modern Library Edition, 1937; 423.

8. Adam Smith, *The Wealth of Nations*, Random House, Modern Library Edition, 1937; 17.

9. Leviticus 25.23.

10. Ched Meyers, *The Biblical Vision of Sabbath Economics*, The Church of the Savior, Washington, DC. 2001; 16.

11. William Greider, *Secrets of the Temple How the Federal Reserve Runs the Country*, Touchstone, New York 1989; 52.

12. Charles P. Kindleberger, *Manias, Panics, Crashes-A History of Financial Crises* Basic Books, USA 1989 in his analysis of financial calamities noted that a lender of last resort, aka corporate bailout/welfare, was needed to prevent markets from collapsing.

13. James Livingston, *Origins of the Federal Reserve System: Money, Class, and Corporate Capitalism, 1890-1913*, Cornell University Press, 1986; 232-233.

14. "Democracy and market economics are not the same thing. Worse, the attempts to confuse and conflate them in pretended equivalence stood out at the millennium as a destructive aspect of U.S. Politics...

In the United States of the turn of the century, the wealth has concentrated with the help of the corruption of politics on one hand the suasion of market idolatry and economic Darwinism on the other...

Ultimately, the guideposts of a market-based society never seem to progress beyond tautology: policies that advance markets are good and efficient because they advance markets. The raw logic of blurring the marketplace and polity, however, boils down to a disturbing simplicity: one dollar, one vote. Inequality is the natural law of the cash driven marketplace. The more you have, the more you can buy. Buying is good. The more you can buy, the more validating your acts.

The next jump is more perverse. Merge politics with the marketplace and buying becomes the game: one dollar, one vote, ten dollars, ten votes." Kevin Phillips, *Wealth and Democracy*, Broadway Books, New York, 2002; 417-419.

15. Ralph Nader,"Cutting Corporate Welfare," http://www.thirdworldtraveler.com/Nader/CutCorp Welfare_Nader.html.

16. William Greider, "Sovereign Corporations", The Nation, April 30, 2001.

17. Wayne Ellwood, "The Great Privatization Grab", New Internationalist, April 2003, Issue 355, http://newint.org /features/2003/04/01/keynote/.

18. Larry Elliot, Will Hutton and Julie Wolf, "Pound Drops Out of ERM", UK Guardian, September 17, 1992.

19. Gregory J. Milkman, *The Vandals' Crown; How Rebel Currency Traders Overthrew the World's Central Banks*, The Free Press, New York, 1995; xi.

20. Graham Searjeant, "Oil speculators put even the masters of Opec in their place", August 20, 2004.

21. Gregory J. Milkman, *The Vandals' Crown; How Rebel Currency Traders Overthrew the World's Central Banks*, The Free Press, New York, 1995; xii.

22. David Hackett Fischer, *The Great Wave-Price Revolutions and the Rhythm of History*, New York Oxford, Oxford University Press; 1996; 251-252.

23. Matthew 6:24.

24. Karl Marx, *Writings of the young Marx on Philosophy and Religion*, Edited and translated by Lloyd D. Easton & Kurt H. Guddat, Doubleday and Company, Garden Center, New York, 1967; 245-246, 268, 289-290.

25. Bank for International Settlements; "Semiannual OTC

derivatives statistics at end-June 2008," http://www.bis.org /statistics/derstats.htm.

26. Without access to a bank.

27. Per the 1998 Survey of Consumer Finance(Federal Reserve): "In 1998, 90.5% of Americans had some type of transactions account—a category comprising checking, savings and money market deposit accounts, money market mutual funds, and call accounts with brokerage firms."; 8.
Stacie Carney and William G. Gale in *Assets for the Poor: The Benefits of Spreading Asset Ownership*, Russell Sage Foundation, 2001, found that 20 percent of American households and 45 percent of black households were un-banked.

28. Unfortunately there is no accurate accounting of the size and scope of the fringe banking market. Carr and Schuetz in "Financial Services in Distressed Communities: Framing the Issue, Finding Solutions", estimated that "fringe banking services had gross revenues of $78 billion in 2000—and nowhere is there an accurate accounting of such." See Madis Senner, "Financial Deregulation—Promoting Discrimination and the Rise of Fringe Banking," http://www.jubileeinitiative.org/RiggedDeregulation.htm.

29. "A manager of PD Chex, a payday lender in Colorado, estimated that only two percent of customers take only one loan...The owner of the store, Avrum Schulzinger, went on to say that he expects all of PD Chex's customers to default eventually'," Consumers Union, "Fact Sheet on Pay Day loan," November 1, 1999.

30. Bhagavad-Gita 6.20-23.

31. William Leach, *Land of Desire: Merchants, Power, and the Rise of a New American Culture*, Pantheon books, New York, 1993; 1.

32. Vance Packard, *The Hidden Persuaders*, Pocket Books, New York 1957; 1-2.

33. Vance Packard, *The Hidden Persuaders*, Pocket Books, New York 1957; 219.

34. Vance Packard, *The Hidden Persuaders*, Pocket Books, New York 1957; 221.
35. Naomi Klein, *No Logo: Taking Aim at the Brand Bullies*, Knopf Canada, 2000; 30.
36. Kalle Lasn, *Culture Jam—How to Reverse America's Suicidal Consumer Binge—And Why We Must*, Quill, New York, 1999; XII-XIII.
37. Alan Greenspan, Panel discussion: "Euro in Wider Circles," at the European Banking Congress 2004, Frankfurt, Germany November 19, 2004.

Chapter 8 An Abundance of Snares

1. John Hagee, *The Seven Secrets*, Charisma House, Lake Mary, Florida; 2004; 210.
2. John Hagee, *The Seven Secrets*, Charisma House, Lake Mary, Florida; 2004; 218.
3. Joel Osteen, *Your Best Life Now*, FaithWords, New York, 2006; 15.
4. Joel Osteen, *Your Best Life Now*, FaithWords, New York, 2006; 52.
5. Joel Osteen, *Your Best Life Now*, FaithWords, New York, 2006; 27.
6. Joel Osteen, *Your Best Life Now*, FaithWords, New York, 2006; 29.
7. Rhonda Byrne, *The Secret*, Atria Books, New York, 2006; 99.
8. Rhonda Byrne, *The Secret*, Atria Books, New York, 2006; .
9. Ched Meyers, *The Biblical Vision of Sabbath Economics*, The Church of the Savior, Washington, DC. 2001; 5.
10. Jeremiah 22.13.
11. Ronald J. Sider, *Rich Christians in an Age of Hunger* Word Publishing, Dallas, 1997; 52.
12. Luke 6.20-21, 24.
13. Mt 19.24.
14. Frances Vaughan, *Shadows of the Sacred; Seeing Through*

Spiritual Illusions, Quest Books, The Theosophical Publishing House, Wheaton, Illinois, USA, 1995; 255.

15. Dana Milbank and Walter Pincus, "Al Qaeda-Hussein Link Dismissed," Washington Post, June 17, 2004.

16. Mike Nizza, "Cheney Unconcerned By Iraq War's Unpopularity," NY Times; March 19, 2008.

17. Moshe Halbertal and Avishai Margalit, *Idolatry,* Harvard University Press, Cambridge Mass, 1992; 127.

18. The Pew Center of the States, "One in 100: Behind Bars in America 2008."

19. The Pew Center of the States, "One in 100: Behind Bars in America 2008."

20. *Agents of Repression; The Cointelpro Papers.*

21. Ward Churchill and Jim Vander Wall, *The Cointelpro Papers, Documents from the FBI's Secret Wars Against Dissent in the United States,* South End Press, Cambridge, Ma., 2002; 1.

22. A 'snitch jacket' is when someone falsely labels a member of a group a snitch, thereby putting their life at risk.

23. United States Senate Select Committee to Study Governmental Operations with Respect to Intelligence Activities, "Intelligence Activities and the Rights of Americans, APRIL 26 (legislative day, April 14), 1976.

24. United States Senate Select Committee to Study Governmental Operations with Respect to Intelligence Activities, "Intelligence Activities and the Rights of Americans, Book 3, Final Report, APRIL 23 (under authority of the order of April 14)," 1976.

25. TRAC (Transactional Records Access Clearinghouse,) "Criminal Terrorism Enforcement Since 9/11," December 8, 2003. http://trac.syr.edu/tracreports/terrorism/report031208. html#figure1. Updated September 9, 2006, http://trac.syr.edu/ tracreports/terrorism/169/.

26. Dan Eggen and Julie Tate, "U.S. Campaign Produces Few Convictions on Terrorism Charges," Washington Post, June

12, 2005.

27. Department of Justice, "INDICTMENTS ALLEGE ILLEGAL FINANCIAL TRANSFERS TO IRAQ; VISA FRAUD INVOLVING ASSISTANCE TO GROUPS THAT ADVOCATE VIOLENCE," February 26, 2003, http://www.usdoj. gov/opa/pr/2003/February/03_crm_119.htm.

28. Renee K. Gadoua, "Up to 150 Questioned; Doctor Denied Bail; Muslims Afraid to Speak Out Publicly," Syracuse Post-Standard, March 1, 2003.

29. Michael Roston, "Former aide: Gonzales, White House counsel signed off on firings," Raw Story, March 29, 2007; http://rawstory.com/news/2007/Leahy_Justice_does_not_ser ve_at_0329.html.

30. Hosea 5.1.

31. Matthew 23.13-39.

32. Matthew 19.30, Matthew 20.16, Mark 10.31.

33. Mark 10.42-45.

34. The Great Law of Peace (Gayanashagowa); no.24.

35. John 13.1-20.

36. James Alison, *Undergoing God, Dispatches from the Scene of a Break-in*, Continuum International Publishing, New York, 2006; 9.

37. Swami Vivekananda, *What Religion Is*, The Julian Press, New York, 1962; 211.

38. Jiddu Krishnamurti, *Freedom from the Known*, edited by M. Lutyens, HarperSanFrancisco 1969; 21.

39. Amos 3.10-11.

40. Rudolph Steiner, *Bees—Lectures by Rudolph Steiner*, Translated by Thomas Braatz, Anthroposophic Press, Great Barrington, Ma, NY 1998; 178.

41. Rudolph Steiner, *Bees—Lectures by Rudolph Steiner*, Translated by Thomas Braatz, Anthroposophic Press, Great Barrington, Ma, NY 1998; 177-178.

42. Rudolph Steiner, *Bees—Lectures by Rudolph Steiner*, Translated

by Thomas Braatz, Anthroposophic Press, Great Barrington, Ma, NY 1998; IX-X.

43. Alexei Barrionuevo, "Bees Vanish, And Scientists Race For Reasons," New York Times; April 26, 2007.

44. Madis Senner, " Wind Turbines Disrupt the Flow of Prana," http://www.jubileeinitiative.org/windmills.html.

45. Christopher Catherwood, *Why Nations Rage, Killing in the Name of God*, Rowman & Littlefield Publishers, Lanham, Maryland 2002; 2.

46. David Bohm, *Wholeness and the Implicit Order*, Routledge, London, 1980; 1.

Chapter 9—Samskaras Can Free Us

1. The Great Law of Peace (Gayanashagowa) no.2.

2. Dhammapada 33.

3. Bhagavad-Gita 6:6.

4. A.C. Bhaktivedanta Swami Prabhupada, *Bhagavad-Gita(As it is)*; The Bhaktivedanta Book Trust, Los Angeles, 1996; commentary to 6:6, 79.

5. Rama Prasad, *Nature's Finer Forces- The Science of Breath & the Philosophy of the Tatwas*, The H.P.B. Press, 1894, http://www.hermetics.org/prasad.html; XIII. Yoga (II), "The Manifestations of Psychic Force."

6. Thich Nhat Hanh, *Creating True Peace*, Free Press, New York, 2003; 18.

7. *The Way of the Pilgrim and the Pilgrim Continues His Way*, translated by Olga Savin, Shambhala, Boston, 2001.

8. 1 Thessalonians 5:17

9. Patanjali, *Yoga Sutras*; 2.33

10. Lama Thubten Yeshe, *Introduction to Tantra, A Vision of Totality*, Wisdom Publications, Boston, Ma, 1987; 48.

11. Bhagavad-Gita 5.24.

12. Mirra Alfassa, The Mother, *The Mother; Collected Works, Volume 9*, Sri Aurobindo Ashram Trust, http://sriau-

robindoashram.info/Default.aspx?1=234, 1972; 415.

13. Mirra Alfassa, The Mother, *The Mother; Collected Works*, Volume 9, Sri Aurobindo Ashram Trust, http://sriaurobindoashram.info/Default.aspx?1=234, 1972: 416.

14. Brihadaranyaka Upanishad 2.4.5.

15. Bhagavad-Gita 8:8.

16. Swami Satyasangananda Saraswati, *Sri Vijnana Bhairava Tantra, The Ascent*, Yoga Publications Trust, Munger, Bihar, India, 2003; 1.

17. A deep meditative state consisting of three steps—dharana, dhyana and Samadhi. It is said that samyama reveals deep or hidden knowledge of an object. *Yoga Sutras*; 3.4-5.

18. Patanjali, *Yoga Sutras*; 3.5.

19. Patanjali, *Yoga Sutras*; 3.18.

20. Patanjali, *Yoga Sutras*; 3:21.

21. Swami Vivekananda, *Raja Yoga*, Ramakrishna-Vivekananda Center, New York; 141.

22. Swami Satyananda Saraswati, *Four Chapters on Freedom— Commentary on the Yoga Sutras of Patanjali*, Yoga Publications Trust, Munger, Bihar, India, 2002; 68.

23. Sogyal Rinpoche, *The Tibetan Book of Living and Dying*, HarperSanFrancisco, 1992; 214-219.

24. Swami Satyananda Saraswati, *Four Chapters on Freedom— Commentary on the Yoga Sutras of Patanjali*, Yoga Publications Trust, Munger, Bihar, India, 2002; 133.

25. Richard Deats, *Mahatma Gandhi Nonviolent Liberator*.

26. Mohandas Karamchand Gandhi, *Satyagraha: Nonviolent Resistance*, Navajivan Publishing House, Ahmedabad, India 1958; 44.

27. Mirra Alfassa, The Mother, *The Mother; Collected Works*, Volume 9, Sri Aurobindo Ashram Trust, http://sriaurobindoashram.info/Default.aspx?1=234, 1972; 417.

28. Thich Nhat Hanh, *The Energy of Prayer, How to Deepen your Spiritual Practice*, Parallax Press, Berkeley, Calif., 2006; 41.

29. Mark 14.22-24.
30. Thich Nhat Hanh, *The Energy of Prayer, How to Deepen your Spiritual Practice*, Parallax Press, Berkeley, Calif., 2006; 93.
31. Thich Nhat Hanh, *The Energy of Prayer, How to Deepen your Spiritual Practice*, Parallax Press, Berkeley, Calif., 2006; 55-56.

Chapter 10 God
1. Matthew 22.36-40.
2. Surah 11.90.
3. 1 John 4.8.
4. Isaiah 54.10.
5. Matthew 5.39.
6. Exodus 22.21-24.
7. Hosea 6:6.
8. Mathew 9:13.
9. Walter Wink, *Engaging the Power, Discernment and Resistance in a World of Domination*, Fortress Press, Minneapolis, MN, 1992; 17.
10. Bhagavad-Gita 4.19-20.
11. Aldous Huxley, *The Perennial Philosophy;* Harper Colophon Books, New York, 1970; 227, 233.
12. Bhagavad-Gita 11.23, 24.
13. Bhagavad-Gita 11.46.
14. Surah 50.16.
15. Matthew 6.5-7.
16. Joseph Epes Brown, *The Sacred Pipe, Black Elk's Account of the Seven Rites of the Oglala Sioux*, University of Oklahoma Press, Norman, Oklahoma, 1953; 43-46.
17. Christopher Isherwood, *Ramakrishna and His Disciples*, Simon and Schuster, New York, 1965; 119.
18. Matthew 25.35, 36, 40.
19. As was noted earlier mantra, like affirmations, ultimately gets you to overcome challenges and believe what you are saying.

20. Psalm 23:4.
21. Bhagavad-Gita 4.10.
22. Bhagavad-Gita 6.14.
23. Bhagavad-Gita 5, 'Karma-yoga—Action in Krishna Consciousness."
24. Bhagavad-Gita 5.10.
25. Bhagavad-Gita 5.12.
26. Several of us from St. Mary's in West Harlem, NY used to 'Ghost' the Federal Reserve in NYC. We divided into groups of two and meet with our partner for prayer before the "ghosting". Each of us would carry a placard, containing scripture and the message we wanted to deliver. We would walk around Wall Street for about and hour in silence and in a contemplative mind. When we were finished we would conclude with a prayer. See: http://www.jubileeinitiative.org/ghost1.html
27. Matthew 18.15-17.
28. Michael L Birkel, *Silence and Witness, The Quaker Tradition*, Orbis Books, Maryknoll, New York, 2004; 105.
29. Gray Cox, *Bearing Witness: Quaker Process and a Culture of Peace*, Pendle Hill Publications, Wallingford, Pa, 1985; 15.
30. Barbara Rossing, *The Rapture Exposed—The Message of Hope in the Book of Revelation*, Westview 2004 Boulder, Co.; 119-120.
31. Walter Wink, *Engaging the Powers, Discernment and Resistance in a World of Domination*, Fortress Press, Minneapolis, Mn, 1992; 89.
32. Page 15; *The Kingdom of God is Within You*; Tolstoy, Leo; Farrar, Straus and Cudahy, 1961
33. Mohandas Karamchand Gandhi, *Satyagraha: Nonviolent Resistance*, Navajivan Publishing House, Ahmedabad, India 1958; 6-7.

Chapter 11 Love and Community

1. Isaiah 53.

2. Mt 20.16.
3. Laotse, Book 2.7, *The Wisdom of Laotse* The modern Library, New York, 1976; 73.
4. Jiddu Krishnamurti, *Commentaries on Living; Third Series from the notebooks of J. Krishnamurti*, Edited by D. Rajagopal, Victor Gollancz, Ltd.; London, 1968; 186.
5. Matthew 22.36-40.
6. "To those who ask about our origin and our founder we reply that we have come in response to Jesus' commands to beat into plowshares the rational swords of conflict and arrogance and to change into pruning hooks those spears that we used to fight with. For we no longer take up the sword against any nation, nor do we learn the art of war any more. Instead of following the traditions that made us "strangers to covenants" (Eph 2:12), we have become sons of peace through Jesus our founder. (Against Celsus 5.33) J. Patout Burns, Robert J. Daly and John Helgeland," John Fortress Press 1985: 39.
7. John Driver, *How Christians Made Peace With* War, Herald Press 1988; 15.
8. Dhammapada 129.
9. Tatvarth Sutra 7.4,11.
10. Isaiah 1.16.
11. Abraham J. Heschel, *The Prophets* Perennial Classics, Harper & Row, New York, 2001; 256-257.
12. Bahaullah, *Gems of Divine Mysteries*.1.
13. 1 Peter 2.24.
14. 1 Peter 2: 21
15. Sogyal Rinpoche, *The Tibetan Book of Living and Dying*, HarperSanFrancisco 1992; 194.
16. Sogyal Rinpoche, *The Tibetan Book of Living and Dying*, HarperSanFrancisco 1992; 193-208.
17. Rig Veda 1.125.5.
18. Matthew 6.3-4.

19. Maxell Taylor Kennedy, *Make Gentle the Life of This World, The Vision of Robert F. Kennedy*, Harcourt Brace & Co., New York, 1998; 66.
20. Dorothy Day, Patrick Jordan Editor, *Writings from Commonweal; Day, Dorothy*, The Liturgical Press, Collegeville, Minnesota, 2002; 57.
21. Dorothy Day, Robert Ellsberg Editor, *By Little and By Little, The Selected Writings of Dorothy Day*, Alfred A. Knopf, 1983; 6.
22. Ronald J. Sider, *Rich Christians in an Age of Hunger* Word Publishing, Dallas, 1997; 49.
23. Matthew 18.19-20.
24. Isa/Isha (Vagasaneyi-Samhita) Upanishad 6-7.
25. Wayne Dyer, *Change your thoughts Change your Life—Living the Wisdom of Tao*, Hay House, Carlsbad, California, 2007; 26.
26. Bahaullah, *The Tabernacle of Unity* 2.34.
27. Michael L. Birkel, *Silence and Witness, The Quaker Tradition*, Orbis Books, Maryknoll, New York, 2004; 105.

Chapter 12 Sacred Earth—Creating Sacred Space

1. Joseph Epes Brown, *The Sacred Pipe, Black Elk's Account of the Seven Rites of the Oglala Sioux*, University of Oklahoma Press, Norman, Oklahoma, 1953; 48.
2. Joseph Epes Brown, *The Sacred Pipe, Black Elk's Account of the Seven Rites of the Oglala Sioux*, University of Oklahoma Press, Norman, Oklahoma, 1953; 48.
3. John Epes Brown notes that "The red Road" is that which runs north and south and is the good or straight way, for to the Sioux the north is purity and the south is the source of life. This "red road" is thus similar to the Christian "straight and narrow." Page 7
4. Joseph Epes Brown, *The Sacred Pipe, Black Elk's Account of the Seven Rites of the Oglala Sioux*, University of Oklahoma Press, Norman, Oklahoma, 1953; 7.
5. "Ithaca's Ceremonial Embers Still Burn,"

http://www.jubileeinitiative.org/SacredIthaca.html.
6. Sang Hae Lee, *Feng Shui: Its Context and Meaning*, University Microfilms International, An Arbor, Mi., 1986; 20.
7. http://www.jubileeinitiative.org/SacredFOL.html
8. Swami Niranjanananda Saraswati, *Prana Pranayama, Prana Vidya*, Bihar School of Yoga, Munger, Bihar, 1998; 8.
9. I often find ceremonial sites located on energy lines or earth chakras. The most sophisticated formation I have found was on South Hill where all the earth chakras are covered with stones. South Hill is located at the southern end of Canandaigua Lake in New York, and is where the Seneca people believe that they were born. I don't think that the Seneca people did this. I believe that it was done by either the Adena or Hopewell because New York State archeologist Ritchie claimed that there were several stone mounds on top of South Hill that he attributed to them. A survey of South Hill in the spring of 2008 with the Finger Lakes Dowsers found that the mounds were placed on energy vortices. A later examination of South Hill showed many of the earth chakras up and down the mountain were similarly covered with stone, energizing the atmosphere as if it were a Nikola Tesla experiment. The rationale for covering the energy vortexes appears to be more spiritual or energizing the people rather than anything to do with healing Mother Earth. See "South Hill the Holy Mountain", http://www.jubileeinitiative.org/sacredsouthhill.html.
10. Swami Vivekananda, *Raja Yoga*, Ramakrishna-Vivekananda Center, New York, 1973; 32.
11. I believe that there once was a civilization in the greater upstate NY area that had knowledge of and could sense consciousness. I have found several places where large stones have been strategically placed in fields of consciousness (www.GaiasSoul.org) that tells me this. I call these people the Spirit Keepers.

12. Chief Jake Swamp, *Giving Thanks—A Native American Good Morning Message*, Chief Jake Swamp, illustrated by Erwin Printup, Jr., Lee and Low Books.
13. Matthew 7.16.
14. Thich Nhat Hanh, *Creating True Peace* Free Press, New York, 2003; 169.
15. http://www.jubileeinitiative.org/sacredKateri.html
16. http://www.gaiassoul.org/
17. "The history of the most important half century of our national life will be imperfectly written if it fails to place Gerrit Smith in the front rank of the men whose influence was most felt in the accomplishment of its results." New York Times, Volume XXIV No. 7265 December 29, 1874. From *Heaven and Peterboro: The Current Relevance of 19th Century Peterboro to Human Rights Today*, Donna Dorrance Burdick, Norman K. Dann, Dot Willsey, Dr. Sheila Johnson Institute, State University of New York, Morrisville, New York, January 2003.
18. Carl Carmer, a folklorist coined the term America's 'psychic highway' to describe all the spiritual fervent/exploration and social justice that was bubbling up in upstate New York during the nineteenth century.

Chapter 13 Heaven on Earth

1. Isaiah 65.25.
2. Matthew 6.10.
3. Srimad Bhagavatam 3.26.6,7.
4. Barbara Rossing, *The Rapture Exposed—The Message of Hope in the Book of Revelation*, Westview 2004, Boulder, Co.; 13.
5. Amos 5.24.
6. 1 John 2.15.
7. Matthew 19.16-22.
8. Everett Gordon, Fredrick Barnes Tolles, *The Witness of William Penn*, The Macmillan Company, New York, 1957. Taken from

Penn's *No Cross, No Crown* (1169, 1682, 1694), "Worship and the Common Life"; 47.
9. Bhagavad-Gita 2.56-59.
10. John 15.18-19.
11. 1 Corinthians 4.16, 1 Corinthians 11.1, Philippians 3.17, 1 Peter 2.21.
12. Luke 6.27.
13. Luke 6.29.
14. Luke 23.34.
15. Mark 10.17-25.
16. Luke 18.25.
17. Mt 5.38-48, Luke 6.27-36.

Chapter 14 Conclusion- Our Faith Can Deliver Us
1. Hosea 2:14-15, 18-20.

Sacred Text Sources

The following versions were used for the notes unless otherwise noted.

Bhagavad Gita, *Bhagavad Gita as It Is*, A. C. Bhaktivedanta, Swami Prabhupada, Bhaktivedanta Book Trust, 1996, c1984. There is also an online version available at, http://www.asitis.com/.

Bible, NRSV.

Quran, *The Meaning of the Holy Quran*, Abdullah Yusuf Ali, Amana Publications, 1999 10th edition, 2001 reprint.

Rig Veda, translated by Ralph Griffith.

Srimad Bhagavatam, translated by Swami Prabhupâda, http://www.srimadbhagavatam.org/contents.html.

Upanishads, translated by Max Muller.

Yoga Sutras, Patanjali. Swami Satyananda Saraswati, *Four Chapters on Freedom — Commentary on the Yoga Sutras of Patanjali*, Yoga Publications Trust, Munger, Bihar, India; 2002.

Selected Bibliography

Alfassa, Mirra: *The Mother; Collected Works*, Sri Aurobindo Ashram Trust.

Aurobindo (Ghose), Sri: *Collected Works*, Sri Aurobindo Ashram Trust.

Besant, Annie and Leadbeater, C. W.: *Thoughts Forms*, Theosophical Publishing House, Wheaton, Ill., 1971, c1925.

Bhagavad-Gita.

Bible.

Brown, Joseph Epes:*The Sacred Pipe, Black Elk's Account of the Seven Rites of the Oglala Sioux*, University of Oklahoma Press, Norman, Oklahoma, 1953.

Burke, James and Ornstein, Robert: *The Axemaker's Gift—A Double Edged History of Human Culture*, G. P. Putnam's Son, New York, 1995.

David-Neel, Alexandra: *With Mystics and Magicians in Tibet*, John Lane The Bodley Head, Ltd, London, England, 1931.

The Great Law of Peace (Gayanashowga.)

Hanh, Thich Nhat: *Creating True Peace*, Free Press, New York, 2003.

Hanh, Thich Nhat: *The Energy of Prayer, How to Deepen your Spiritual Practice*, Parallax Press, Berkeley, Calif., 2006.

Le Bon, Gustave: *The Crowd: A study of the Popular Mind*, Dover Publications, Mineola, NY, 2002 (first 1895, English 1896).

Quran.

Rinpoche, Sogyal: *The Tibetan Book of Living and Dying*, HarperSanFrancisco, 1992.

Suzuki, David: *The Sacred Balance—Rediscovering Our Place in Nature*, Prometheus Books, Amherst, New York, 1998.

Saraswati, Swami Satyasangananda: *Sri Vijnana Bhairava Tantra, The Ascent*, Yoga Publications Trust, Munger, Bihar, India, 2003.

Saraswati, Swami Satyananda: *Four Chapters on Freedom—Commentary on the Yoga Sutras of Patanjali*, Yoga Publications Trust, Munger, Bihar, India, 2002.

Upanishads.

Vedas.

Vivekananda, Swami: *Raja Yoga*, Ramakrishna-Vivekananda Center, New York, 1973.

Wink, Swami: *Engaging the Power, Discernment and Resistance in a World of Domination*, Fortress Press, Minneapolis, MN, 1992.

Lama Thubten Yeshe, *Introduction to Tantra, A Vision of Totality*, Wisdom Publications, Boston, MA, 1987.

Recommended Reading List

The following are books and websites for further learning.

Mother Earth
Mother Earth Prayers: www.MotherEarthPrayers.org.
Gaia's Soul: www.GaiasSoul.org.
Geographic Samskaras:
www.jubileeinitiative.org/Samskaras.html.
Mid-Atlantic Geomancy: www.geomancy.org/.

Mindfulness
Thich Nhat Hanh has written several good books on controlling
your thoughts; *The Miracle of Mindfulness* and *Peace Is Every Step:
The Path of Mindfulness in Everyday Life* are two exceptional ones.
Wayne Dyer, *The Power of Intention*.
Eckhart Tolle, *The Power of Now: A Guide to Spiritual
Enlightenment*.

Vedanta
Bhagavad-Gita.
Ramakrishna-Vivekananda Center of New York;
www.ramakrishna.org/.
Vedanta Society of New York: www.vedantany.org/.
Vedanta Society of Southern California; www.vedanta.org/.

Yoga and Patjanli
Patanjali has been called the father of yoga and his *Yoga Sutras* is
a classic well worth reading. Many people consider the body
exercises taught in the west as yoga, they are asana, the third
limb of the eight limbs of yoga. Each limb is meant to prepare us
to be better able to meditate for long periods so that we may
connect with the divine.

Prabhavananda and Christopher Isherwood, *How to Know God: The Yoga Aphorisms of Patanjali,* Swami. This is a good primer for people wanting to know more about the philosophy of yoga and Vedanta.

Swami Vivekananda, *Raja Yoga.* This is an insightful look at Patanjali from one of the great mystics. This is the only book that Swami Vivekananda wrote.

Swami Satyananda Saraswati, *Four Chapters on Freedom— Commentary on the Yoga Sutras of Patanjali.* This is a great technical resource for anyone that is looking to attain Samadhi via concentration. This is not for beginners.

The Bihar School's Yoga Magazine has lots of information and a search function that you can use to find all sorts of information: www.yogamag.net/.

Tantra
Hindu: Swami Satyasangananda Saraswati, *Sri Vijnana Bhairava Tantra, The Ascent.*

Tibetan Buddhism: Lama Thubten Yeshe, Phillip Glass, Jonathan Landaw, *Introduction to Tantra, A Vision of Totality.*

The Mystical World
Annie Besant and C. W. Leadbeater, *Thoughts Forms.*

Michael Talbot *The Holographic Universe.*

The Sri Aurobindo Ashram Trust contains the Collected works of Sri Aurobindo, the Mother and much more: www.vedantany.org/

Dowsing
The following dowsing organizations have listings of dowsing groups in their respective countries, contain literature, sell dowsing products and books and host conferences.

American Society of Dowsers: http://www.dowsers.org/.

British Society of Dowsers: http://www.britishdowsers.org/.

Canadian Society of Dowsers: www.canadiandowsers.org/.

Canadian Society of Questers: www.questers.ca/.
Listing of International Dowsing Groups:
www.britishdowsers.org/whats_on/international_dowsing
_group.shtml
Sig Lonegren's *Spiritual Dowsing* is a good primer to learn the
basics of dowsing.

Renunciation
Herman Hesse's novel *Siddhartha* tells the story of a spiritual
quest of young man living at the time of Buddha.
Atma Jyoti Ashram, Monastic life:
www.atmajyoti.org/monastic.asp.
Thanissaro Bhikkhu, "Trading Candy for Gold Renunciation as a
Skill";
www.accesstoinsight.org/lib/authors/thanissaro/candy.html.
T. Prince, "Renunciation":
www.accesstoinsight.org/lib/authors/prince/bl036.html.

Index

B O O K S

O is a symbol of the world, of oneness and unity. In different cultures it also means the "eye," symbolizing knowledge and insight. We aim to publish books that are accessible, constructive and that challenge accepted opinion, both that of academia and the "moral majority."

Our books are available in all good English language bookstores worldwide. If you don't see the book on the shelves ask the bookstore to order it for you, quoting the ISBN number and title. Alternatively you can order online (all major online retail sites carry our titles) or contact the distributor in the relevant country, listed on the copyright page.

See our website **www.o-books.net** for a full list of over 500 titles, growing by 100 a year.

And tune in to myspiritradio.com for our book review radio show, hosted by June-Elleni Laine, where you can listen to the authors discussing their books.

MYSPIRITRADIO